WASHINGTON STATE

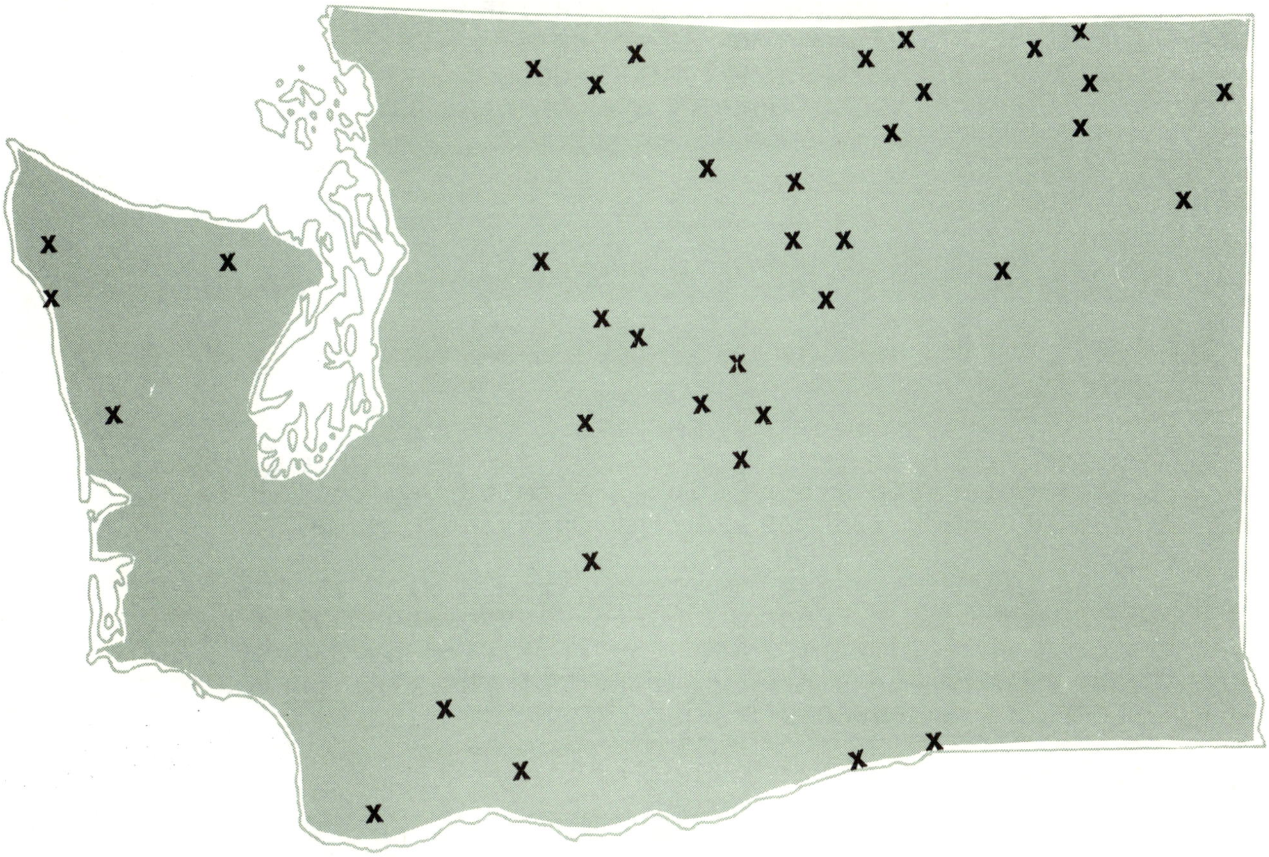

GOLD MINES

ROY F. MAYO

OTHER BOOKS
BY
ROY F. MAYO

GOLD AND STRYCHNINE
GIVE ME LIBERTY
SALMON RIVER GOLD
"GOLD" PAN IT YOU'LL LIKE IT

CREDITS

Topographic maps are reprinted from US Geological Survey 1:100,000 and 1:250,000 series Quadrangle Maps.

Mining Claim maps are reprinted from MINING IN THE PACIFIC NORTHWEST. L. K. Hodges, editor 1897. Published by THE SEATTLE POST-INTELLIGENCER and reprinted by SHOREY'S BOOK STORE, 1967. Reprinted courtesy of THE SEATTLE POST-INTELLIGENCER and SHOREY'S BOOK STORE.

Illustrations by Bert McWhorter.

Printed and bound in the United States of America.

Printed By:
CLASSIC PRINTING
201 E. MEEKER • KENT, WA 98032

Published by

NUGGET ENTERPRISES
P.O. Box 184
ENUMCLAW, WA. 98022

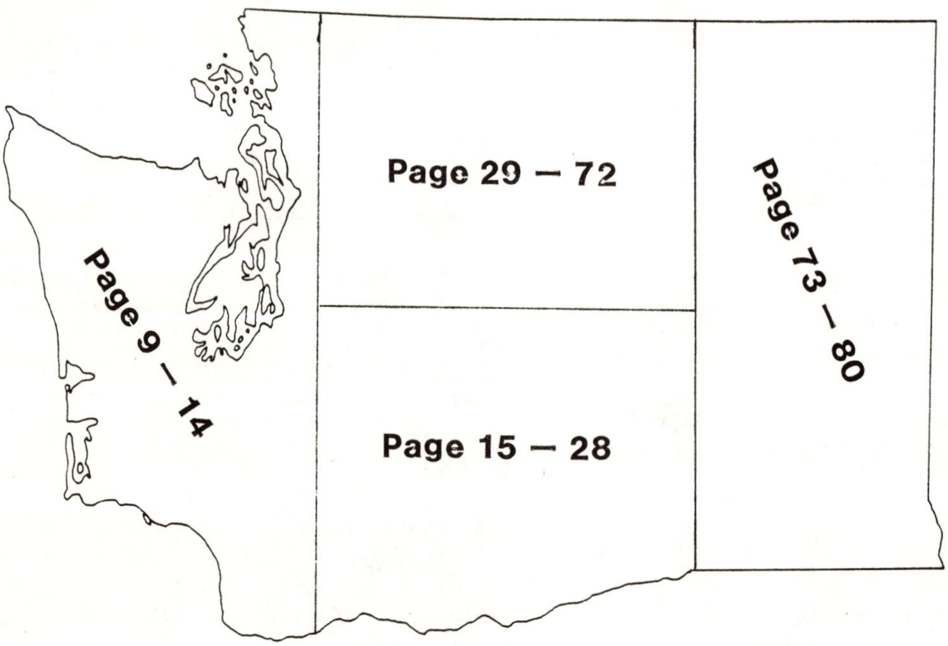

Page 29 — 72
Page 9 — 14
Page 73 — 80
Page 15 — 28

WASHINGTON STATE GOLD MINES

Miners in Washington State have produced an estimated 3,000,000 ounces of gold since the first discovery in 1852. This includes both lode and placer. Many experts believe that only a small amount of the gold has been mined. This means that only the easy to find gold has been recovered.

Huge amounts of gold remain in the earth. Many known deposits are being held in reserve by companies waiting for better prices or other reasons. Some low grade deposits are not profitable to work using the methods available today.

Most of the old mining areas still have some active miners or prospectors working. Most land in these old areas is under claim. Old mines are being reopened and new discoveries have been made recently. The gold is there. Hard work, persistence and knowledge will find it.

The old saying that if you are looking for gold, you should go where gold has been found, is good advice. However, there is another saying, "gold is where you find it". There have been several discoveries in recent years far removed from the old established areas. Some of these can never be worked, others show promise.

Gold was found in a deep test hole drilled for the pier foundations for the Vantage bridge over the Columbia River. Gold was found in a drainage dry well near Buckley. Gold was found in quartz in the Green River near Black Diamond. All these are far from any mining areas.

An unusual discovery was made by a homeowner near Randall. While cleaning volcanic ash off his roof with a garden hose he found several flakes of gold in his gutters. Was this gold deposited by bird droppings or did Mt. St. Helens distribute gold throughout the ashfall area?

Most counties in Washington State have produced some gold. With the northern half of the State being the most productive. The first reported discovery was made by a party of surveyors looking for a railroad pass through the Cascade Mountains in 1852. The exact location is not clear, but it is believed to have been on the Yakima River.

Within twenty years several major discoveries had been made. Miners stampeding to the Canadian mines found gold on the Similkameen River in 1859. Gold was found on Swauk Creek in 1868 and mining has been continuous ever since. Most gold produced before 1900 was from placer mines, with lode mines being the leader since that time.

The exact value of gold production will never be known. Early reports are only estimates. Miners then and now were and are very secretive about the value of their claims. With the ever present danger of robbery, most miners buried their gold for safe keeping. A miner on a good claim would say he was barely breaking even. Others with poor claims would exaggerate their value hoping to sell to the first sucker with a pocket full of cash.

A considerable amount of gold has been mined without benefit of staking a claim. A prospector finding a small pocket might prefer to work in secrecy rather than have a registered claim showing his location. Others might consider the inconvenience and red tape more trouble than it was worth.

With the dollar value of gold at a price undreamed of a few years ago, interest in gold mining is at an all time high. For many years the price of gold was $20.67 per ounce. In 1934, gold was withdrawn from private ownership and the price was raised to $35.00 per ounce. This fixed price held for another thirty five years, then was raised through several changes to its present official price of $42.22 per ounce.

At the same time the unofficial price was allowed to float at world market price. American citizens were also allowed to own gold. Since then the laws of supply and demand have driven the price to a record $850.00 per ounce. The price has fluctuated considerably since, with a low of $290.00 in the summer of 1982, and a general upswing since.

With these high gold prices, it is now profitable to work lower grade ores and placers. Ground that has lain dormant for many years is now being retested and explored. Almost all ground in established mining areas is now under claim.

This book is not intended to tell you where you can find gold. It is meant to tell you where gold has been found. Most of these locations are proven as to value and are shown in public records. Many are from my own experience.

My primary source of information has been from publications by the State of Washington, Department of Natural Resources, Division of Geology and Earth Resources. This department has a wealth of valuable information of interest to prospectors for gold and other minerals in this state.

An excellent source of historic information is "Mining in the Pacific Northwest" by L.K. Hodges, published by the Seattle Post-Intelligencer in 1897 and reprinted by Shorey Book Store. This book describes and shows maps of virtually every mine or prospect worked in the first twenty-five years of the mining industry in Washington State.

Much of the current interest in gold mining is in the romance of finding gold. Mining, at its best, is a risky and dangerous business. For the newcomer, it soon becomes apparent that most of the romance is connected to the handle of a pick or shovel.

Finding a place to work can sometimes be a problem. All proven mining areas have active claims. Many claim holders believe that their claim has $1,000,000.00 in gold on it. I have known some of these men many years and their stated values have not changed over the years. Some of these owners are willing to protect their million dollars with their lives if necessary, but perferably with yours.

Trespass laws must be observed in all cases. The claim holder owns all minerals on his claim. Some claim owners will allow you to prospect on their claim, provided you show him where you worked and what you found. Others are willing to lease their claim on a percentage basis. A serious prospector can always find a claim to work.

Some mining areas are now within the boundaries of National Parks or Wilderness Areas. In most cases these are now closed to prospecting or mining.

The birthplace of Washington State's mineral industry was on the Similkameen River in 1859. Surveyors mapping the boundary between the United States and Canada are credited with making this discovery. The rush was on. Hundreds of men swarmed into the area. News of rich discoveries in Canada sent the stampeders north and the Similkameen was abandoned for a time.

Many of these stampeders failed to find their bonanza and returned to Washington Territory. At the same time more men were coming into the area from the California mines. Many of these men were inexperienced miners and were only looking for placer gold. Placer gold is easy to recognize and simple to recover. Very little knowledge is required to shovel gravel into a sluice box.

These early miners were of a restless nature. The report of a new strike would set off a new stampede. Good diggings would be left, hoping to find better somewhere else. Most of these men were not lazy. It is almost unbelievable the amount of work that was accomplished by hand methods only.

As the easily accessible placers were worked out, the miners moved upstream into the rugged mountains. Most of the lode mines were located by placer miners tracing colors upstream. The more experienced mineral men recognized minerals other than gold. These men discovered the rich lodes of silver and copper and other metals.

Some lode mines were accidentally discovered by inexperienced miners. An outcropping of beautiful, shiny pyrite crystals was found north of Mt. St. Helens by three hunters. They loaded their packs with all the fools gold they could push in and headed for town with their bonanza. There was some gold in their pyrites but their discovery set off a rush to stake copper claims.

Some of these early lode discoveries were very rich in free gold in the surface outcrops. Other minerals will leach out leaving a high concentration of gold. As these mines were explored to depth, the gold values declined and were no longer profitable to work. Many of these mines were later developed into silver or copper mines.

Development of mines in the higher elevations of the Cascade Mountains was very difficult. Roads and trails were nonexistent. Trails built one year could be destroyed by winter snow avalanches or spring floods. Mine improvements and buildings were often carried away by snow slides. Miners returning in the spring sometimes found their tunnels or shafts covered and filled by rock slides.

All these natural obstacles and discouragements were overcome. The mines were developed and ore produced. Pack trails were built. Wagon roads were built, even railroads. Six smelters were built between 1887 and 1905 to process the production from these booming mines. Only one of these has survived to the present time. The Tacoma smelter, built in 1890 is still operating, but now processing mostly foreign ore.

Mining and mining methods have changed drastically over the past century. The romantic life of the jackass prospector is now gone forever. His tall tales are all in the history books. He found the rich placers and the big nuggets. If the nugget wasn't big enough to pick up with his fingers he would not worry with it. Gold dust was hard to hold in the sluice box and too time consuming to separate from the black sand.

He checked out the outcroppings of quartz. In his search for nuggets, quartz float in the stream bed led him to the source. If the quartz carried enough gold to be seen with the naked eye he was interested, if not he left it for better prospects elsewhere.

Many of these early discoveries were impractical to work. Roads and trails were nonexistent. Processing plants and smelters were hundreds of miles away. Some ore from early mines had to be shipped to Denver or Saint Louis for processing. Ore that would be considered bonanza grade today was worthless under those conditions.

Over the years processing and concentrating plants were built at or near many mines. Many miners built arrastras. This primative grinding mill was simple to build and could be powered by man, mule, or water. These were capable of grinding a ton of ore a day.

The stamp mills were a great improvement. They had their day and were replaced by jaw crushers, ball mills, and roller mills. The Pelton water wheel was a great improvement over the cumbersome undershot and overshot water wheels.

Steam, gasoline, and electric power had their turn. Air operated drills replaced the single jack. The dangerous black powder was replaced by safer and more efficient explosives. Mucking machines replaced the pick, shovel, and wheelbarrow. Buldozers rip apart entire mountains in open pit mining.

Science and technology have now replaced the old prospectors eyeball assay. Satellites now send information from outer space on where to prospect here on earth. In using geochemical prospecting methods, it is possible to determine from a soil sample what minerals are under ground. By geobotanical methods you can determine what minerals are underground by looking at the plants growing on the surface.

Electronics have a big role in today's prospecting. Metal detectors can now tell if a rock has metal in it and in many cases can tell what the metal is. These machines can penetrate through ferrous rock and pin point a gold or silver nugget behind it. It is no longer necessary to crush and pan an ore smmple to determine mineral content. One pass with a metal detector and you have your answer.

The gold pan, as a mining tool, is as useful and necessary today as it was to the 49ers. The modern versions of gold pans are almost foolproof in their ability to trap and hold even the finest particle of gold. The early designs, made of wooden bowls, woven willow baskets, and steel pans with riveted joints, were satisfactory in their time. The early miner had no interest in the fine particles we must save today.

The sluice has also evolved from the primative log and rock riffles to designs that will hold fine particles of gold almost completely cleaned of black sand. The most efficient designs used by the early miners were capable of recovering only the coarse gold. Black sand clogged the riffles and most of the fine gold was lost. These old sluice box locations may be a rich source of gold today.

Even with all the modern advances in mining methods, there are occasions when the old methods still work best. Buldozers can move enough ore to make very low grade deposits profitable to work, but they leave an irrepairable scar on the mountain. Underground work by hand leaves little evidence of your work. In many areas old river channel may be found far up the mountainside above the present stream. These may have many feet of overburden hiding the bedrock. The most efficient methods of working these deposits would destroy the landscape forever. The old pick, shovel, and wheelbarrow drifting methods are the only practical solution to reaching these riches.

Research can be a valuable aid in today's prospecting. A study of claim recordings may reveal the location of an unworked section on a placer stream. Many of the old mining areas have local histories written. These may be found in local libraries or historical societies.

Many books have been written on how to pan gold. Most of these are directed to the recreational prospector. These are available at most book stores or mining equipment suppliers. More technical subjects on specialized mining techniques may be ordered from suppliers such as "Miners", Riggins, Idaho or Pearl Electronics, Seattle, Washington.

The library of the Division of Geology and Earth Resources has a file on most of the mines that have been worked in Washington State.

Shorey's Book Store in Seattle, Washington has a special department dealing with books on mining and prospecting. Shorey's have reprinted several old books of interest to prospectors. Outstanding examples of these are:

"Mining in the Pacific Northwest", published in 1897.

"A Broad Bold Ledge of Gold; Historical Facts of Monte Cristo", published in 1958.

"The Prospector's Manual", published in 1905. This is an excellent source of information on how mining was done in the early days. Some of this practical information is as useful today as it was almost eighty years ago.

Maps are a valuable aid in prospecting today. With the use of topographic quadrangle maps it is possible to locate favorable stream conditions for placer mining. You can determine the easiest access into remote areas. Base maps used in this book are U. S. Geological Survey 1:100 000 scale Topographic Quadrangles. These are metric scale of approximately 5/8 inch per mile. All locations described in this book are also referenced to the 15' (one inch per mile) or 7½' (two and one half inch per mile) quadrangles for greater study details. Lode mining areas and placer streams are shown by red lines on base map.

Each generation of miners in Washington State has recovered its share of this great natural wealth. Each generation has searched for the best place to work. The early miners stampeded from one promising place to the next, always looking for the "best place".

Your share of gold is out there. With the use of this book you may find your "best place". Patience and hard work will produce gold. The pleasures of working with nature in the great outdoors is a golden experience in itself. Good luck and may all your pans be filled with NUGGETS.

PACIFIC OCEAN BEACHES

The ocean beaches along the coast of Washington have produced considerable gold. This is usually a very fine gold. Beach erosion and reconcentration by storm wave action should replenish this source of gold periodically. The most productive area was from Cape Flattery south to La Push. These areas are shown on Cape Flattery, Ozette Lake and La Push Quadrangles in Clallam County.

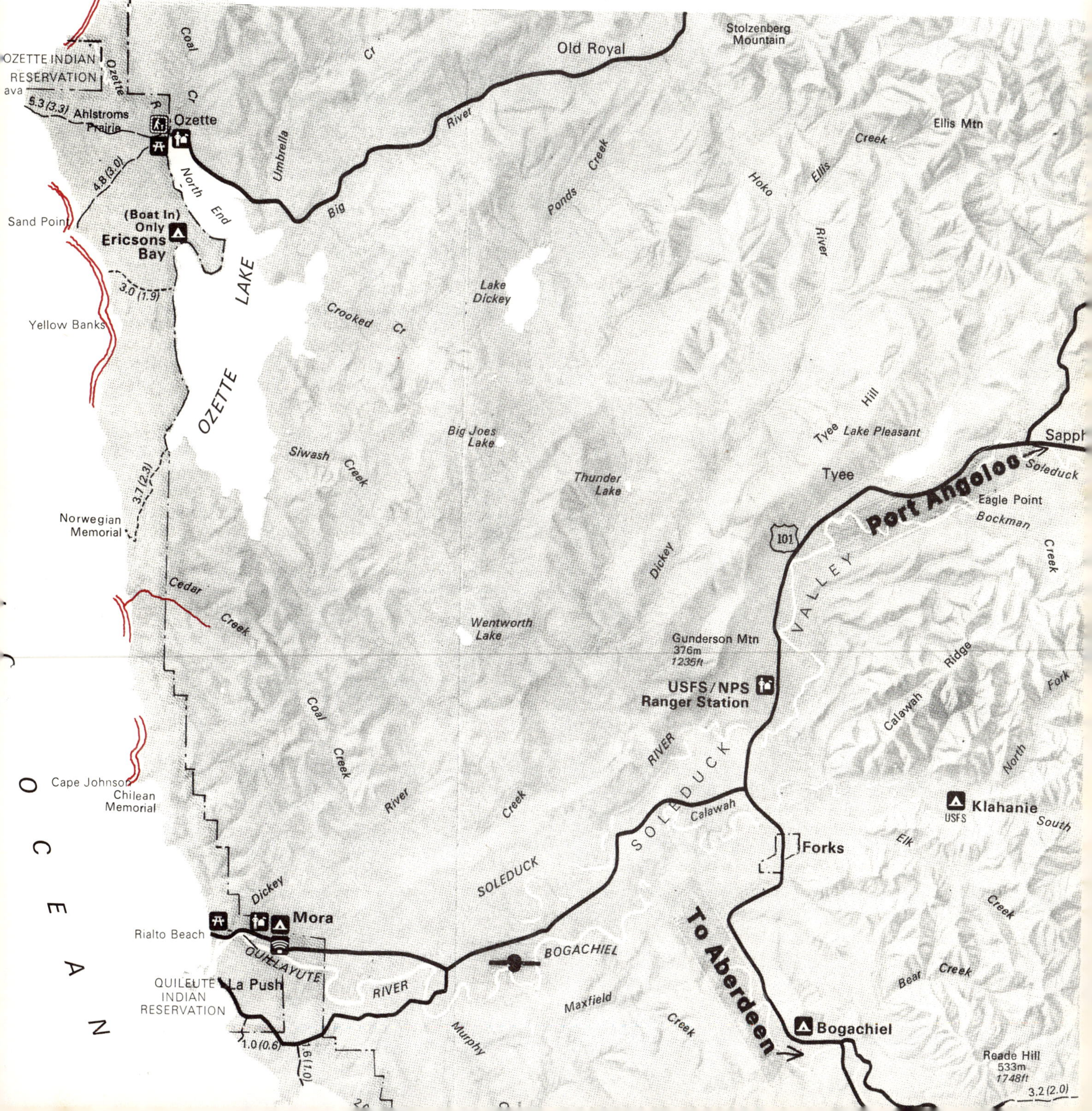

Ruby Beach in Jefferson County has very fine gold in the beach sand. Destruction Island Quadrangle shows this area.

The Moclips, Copalis Beach, and Point Brown Quadrangles in Grays Harbor County show the famous Razor Clam Beaches. Fine gold may also be found here.

Sand bars at the mouth of the Columbia River have produced some gold. The beach at Ocean Park has also produced. Ocean Park, Cape Disappointment and Chinook Quadrangles cover this area of Pacific County.

Vast quantities of sand would have to be processed to make these beach placers profitable. Present regulations prohibit any commercial operation, but a recreational gold panner may find a few colors.

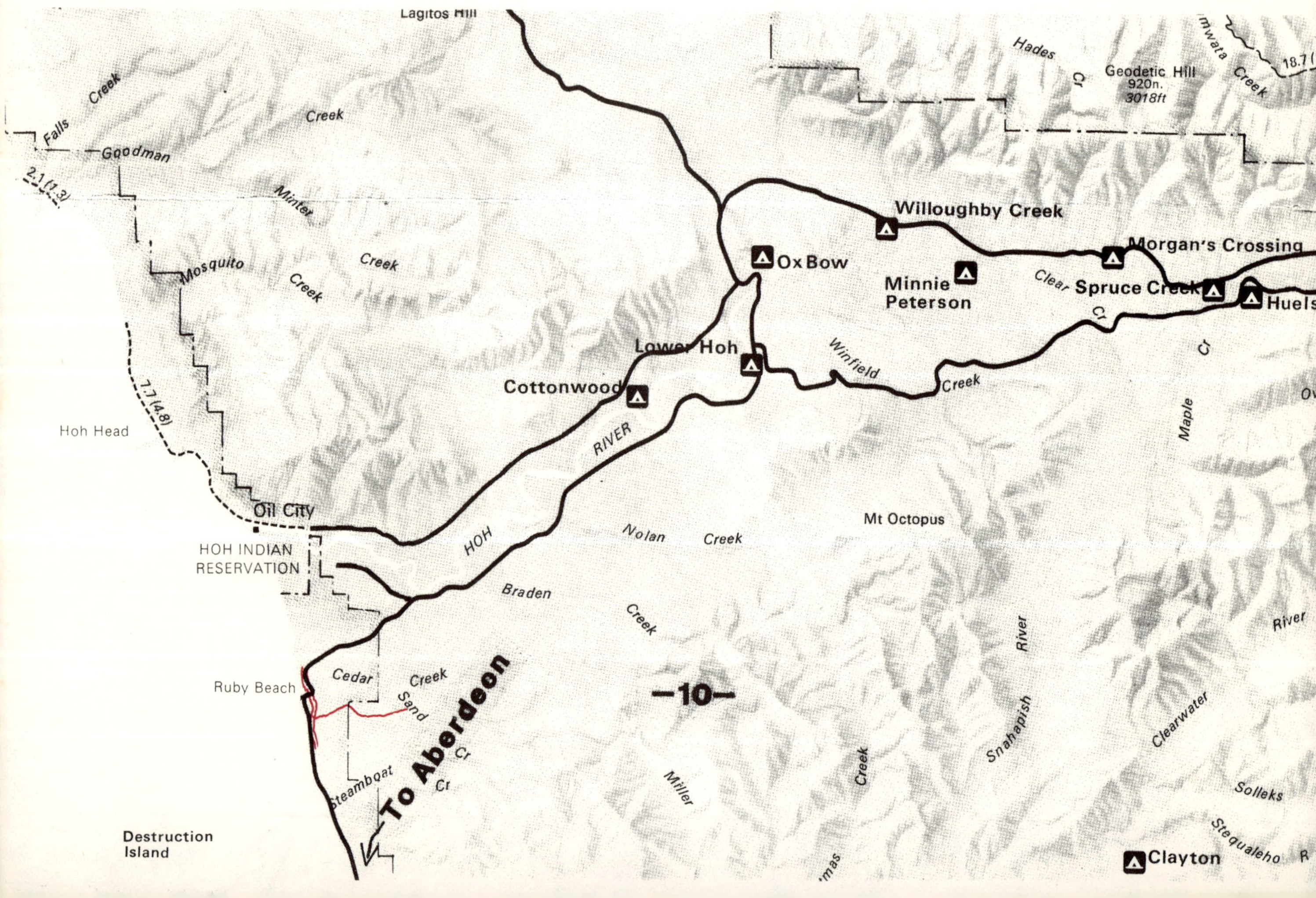

THE COLUMBIA RIVER

The Columbia River drains gold rich areas in five states and Canada. Some of the richest mining areas in the Northwest are on the headwaters of the Columbia River or its tributaries. Floods on these river systems over the ages have created many rich gold placers.

Most of the richest placers are covered by lakes formed by the numerous dams across the river. These are now lost forever to all forms of mining except diving and dredging.

The Lower Columbia River has produced very little gold. Any gold carried that distance from its source would be microscopic in size. Power dams built on the Columbia and many of its tributaries have effectively stopped any future deposits of even microscopic gold except in areas of local enrichment.

Islands between Alderdale and Patterson in Benton County have produced some flour gold. Old river channels in this area may also produce gold. Recently skin divers operating a suction dredge claim to have found coarse gold and small nuggets. A study of river navigation charts may reveal the location of old river channel or sand bars. This area is shown on Alderdale, Crow Butte, Boardman, Blalock Island, and Patterson Quadrangles.

Chinaman Bar on the north bank of the Columbia River between Priest Rapids Dam and Vernita was worked extensively for many years. Evidence of placering can still be seen. This bar is not being replenished by floods because of the dam just upstream, but colors can still be found on the bar and on the beaches. Priest Rapids 15' Quadrangle shows this area of Clark County.

From the vicinity of Vantage north to the Canadian border, placer gold has been found in paying quantities on almost every bar. There were many temporary camps. Chinese miners had the patience to recover this fine gold dust. The amount of gold recovered can only be estimated, but it is believed to be considerable.

This section of the Columbia River is now almost a continuous lake, with very few sections of free flowing water. All these bars are now flooded forever. Divers with dredges may still find a few profitable locations.

All the early activity along the river has left us with several lost treasure legends. Most miners buried their gold for safekeeping, especially the Chinese. If he was accidentally killed or murdered, his gold is still where he left it.

All miners in a Chinese camp near Wenatchee were massacred. Another camp near Trinidad suffered the same fate. Their gold is in rotted buckskin sacks under a rock or in a crack in the basalt canyon wall, but under water.

A rich treasure in the Trinidad area may still be found. A group of miners returning from the Canadian mines saw the Chinese camp being attacked by Indians. They hid on a ledge high up on the canyon wall. Fearing for their lives they buried their gold and most of their supplies. They escaped safely, but did not return for many years and then could not locate the cache. The ledge where they hid is now under water, but many years later their pack saddles were found in a side canyon above the lake level.

This once isolated and primative area is now developed into very productive fruit orchards. West Bar and Babcock Ridge 7½' Quadrangles show this part of Douglas County.

THE SNAKE RIVER

The Snake River has suffered the same fate as the Columbia River. It is now a quiet water lake for almost its entire length in Washington State. Boats and barges can now travel from the Pacific Ocean to Lewiston, Idaho with hardly a ripple on the water.

Flour gold has been found on several bars in small quantities. Washington State University tested several of these bars to determine the most effective method of recovering these fine particles. Their engineering department designed a low cost machine which was used in small commercial operations near Clarkston, Asotin, and Penawawa. These locations are all now under water.

Lost treasure might be the best source of gold on this river. Many miners returning from the mines in Idaho floated down the Snake River by boat or on a log raft. At night they camped on a sand bar. As was customary, each night their gold was hidden or buried. If they did not survive the night, their gold is still there.

WESTERN WASHINGTON

Discoveries of gold have been rare west of Interstate Highway 5. The ocean beach placers have probably produced the most gold. Sand and gravel operations have produced a small amount of gold. Sluice boxes set up to process the gravel wash water have recovered some gold in areas of glacier deposits.

The world's greatest reserves of gold are in the water of the oceans. The amount of gold in sea water is almost unlimited. An attempt to recover this gold was made at Lakota on Puget Sound during the 1930's. Their tests were successful in the laboratory, but a plant built on a pier failed to recover the gold economically enough to pay expenses.

Several place names refer to gold, silver, money, or fortunes. Some of these may refer to actual discoveries, others to wishful thinking and some to plain sarcasm. Jefferson County has its Gold Creek. This is in a mineralized area of manganese and copper on Iron Mountain. Hundreds of feet of tunnel was dug to develop the Tubal-Cain Mine and others. Apparently the hoped for riches did not prove out. There is no hint of gold here, but Gold Creek may be located on the Tyler Peak Quadrangle.

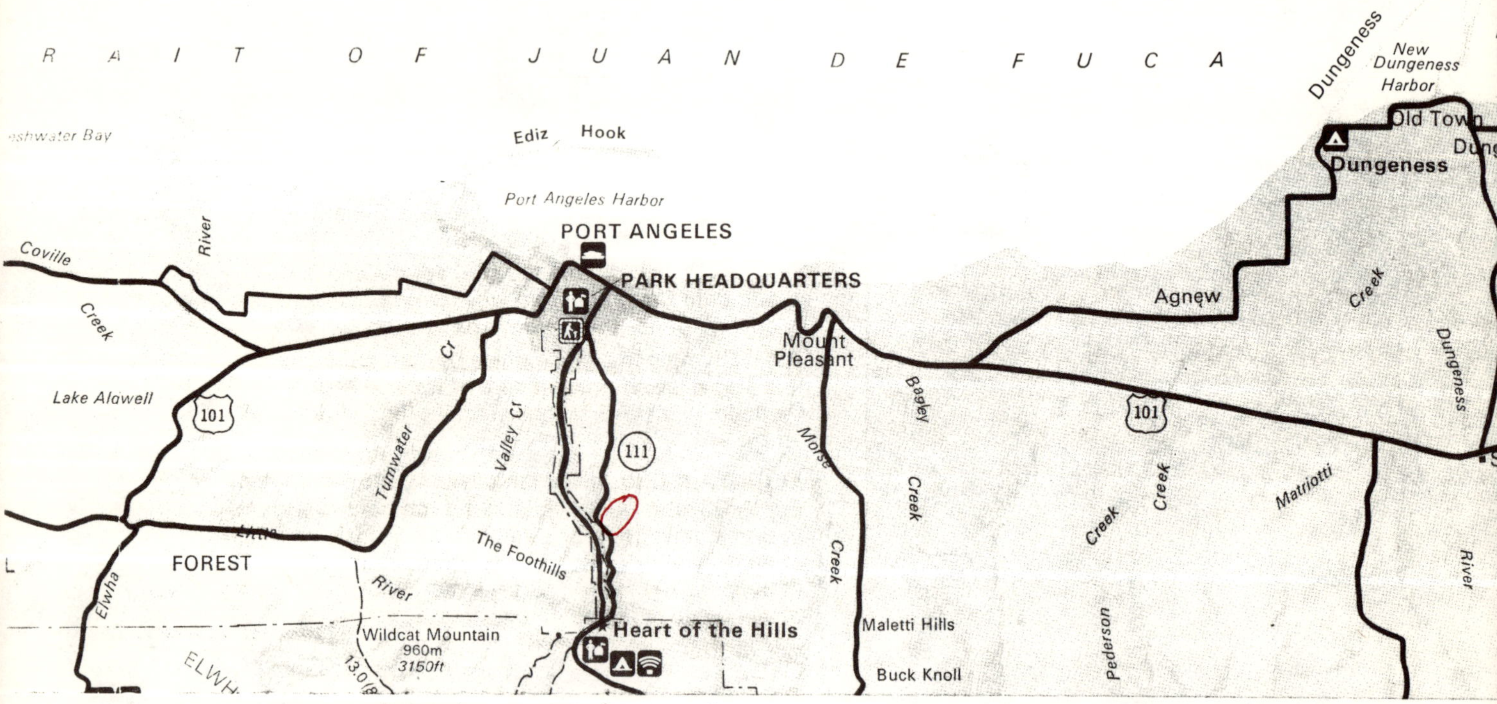

Quartz carrying gold was found between Ennis Creek and White Creek, two miles south of Port Angeles. This area is shown on Port Angeles Quadrangle.

SOUTH CENTRAL

This area of the Cascade Mountains has a few scattered discoveries of gold and some interesting Lost Mine prospects. Some of the first verified discoveries were probably made while searching for the Lost Mine of the Klikitat Indians. These Indians traded gold to the fur traders and the early settlers along the Columbia River. This mine may be located near the mouth of Black Canyon on the Little Klikitat River. Goldendale Quadrangle shows this area of Klikitat County.

The Spaniard's Lost Gold Mine is supposed to be on the slopes of Mt. St. Helens. Some clues indicate that this may be the same as the Lost Klikitat Mine, although the two areas are many miles apart. The skeletal remains of a man and a mule, along with prospecting gear, was found on Mt. St. Helens. This was thought to be the Spaniard, but his mine was never found. The Mt. St. Helens Quadrangle might give a lead.

The St. Helens Mining District has some of the richest mines in the Southern Cascades. They should be rich. Enough money has been poured into them to retire on. As a matter of fact, most of these mines retired in the early 1900's and Mt. St. Helens has mercifully given them a decent burial.

Thousands of feet of tunnels and shaft have been dug in an effort to develop the copper, silver, and gold prospects in this area. The Sweden Mine alone invested $700,000.00 and sold $286.00 of ore in 1905 and $1,202.00 in 1927. Not a very good return.

Small amounts of placer gold has been found on the Toutle River and many of the creeks throughout the area. Most of this area is shown on the Spirit Lake Quadrangle.

Chinese miners had a sluicing operation on Burnt Bridge Creek near 39th Street in Vancouver. The south bound lane of I-5 covers their dam and holding pond. They sluiced an estimated forty to fifty thousand yards of gravel from the hillside north of I-5 highway.

Placer gold has also been found on Brush Prairie, north of Vancouver. Portland 15' Quadrangle shows these two locations.

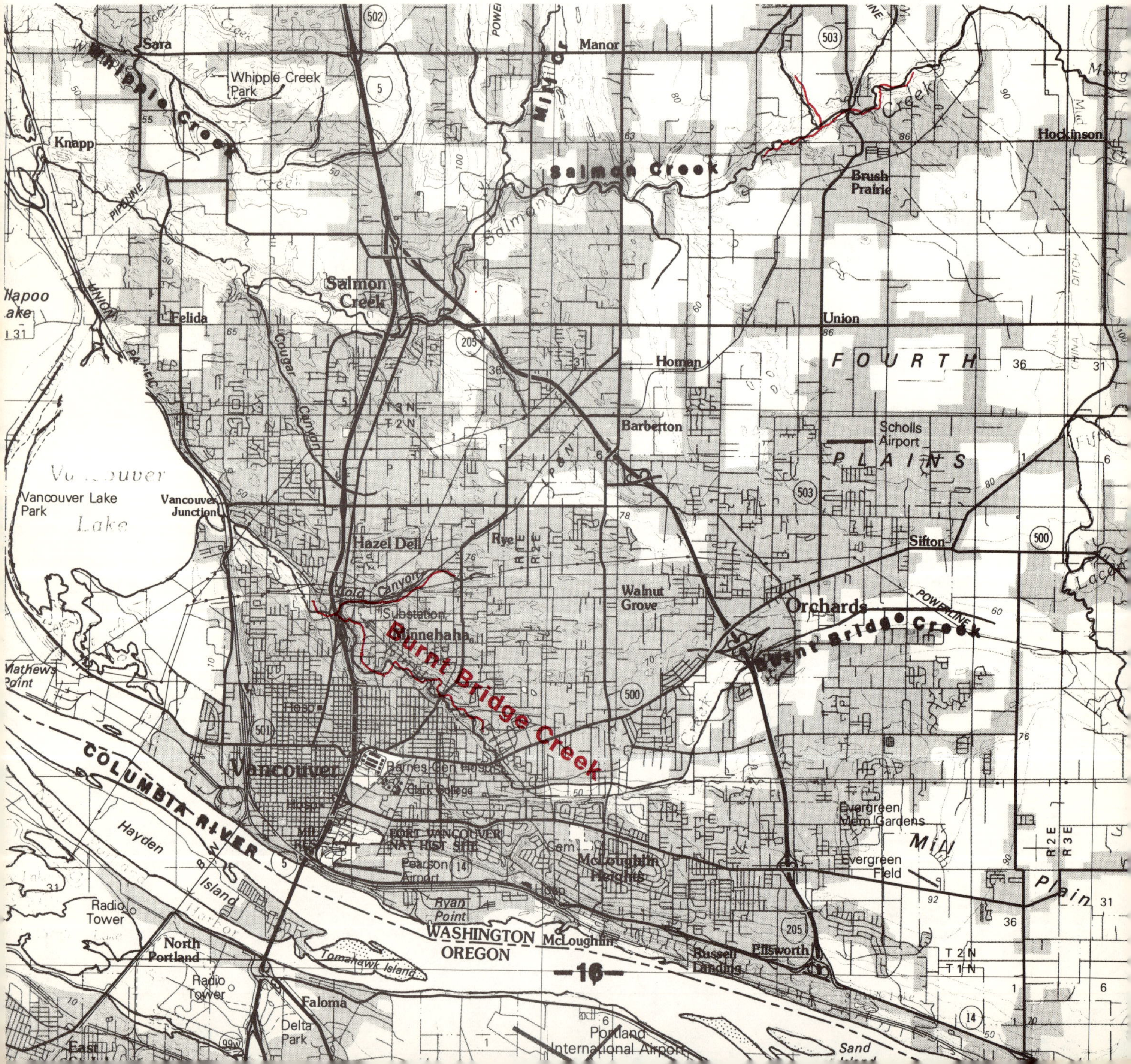

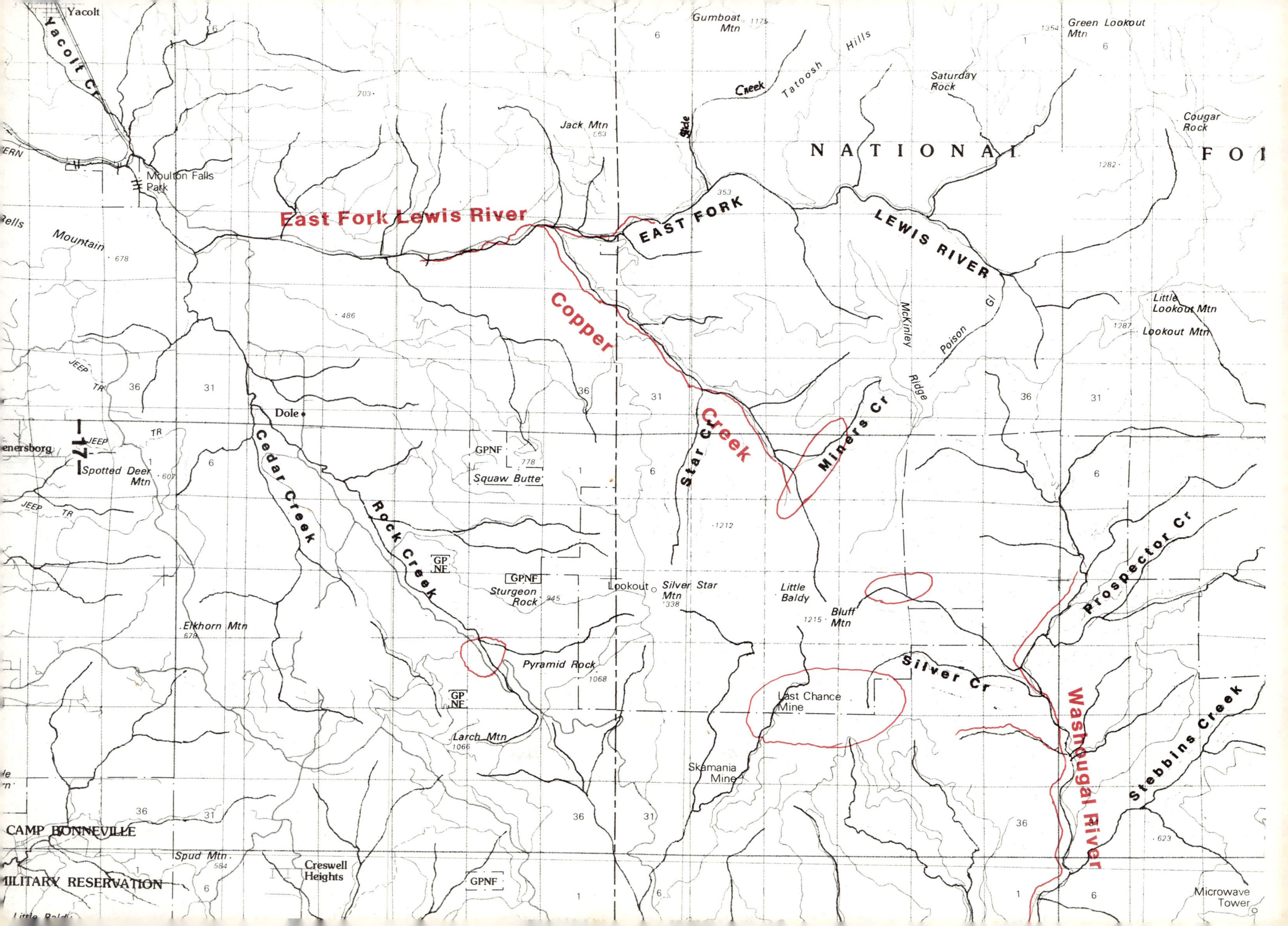

Yacolt
Yacolt Cr
Moulton Falls Park
Mountain
East Fork Lewis River
EAST FORK
LEWIS RIVER
NATIONAL
Gumboat Mtn
Jack Mtn
Tatoosh Hills
Creek
Saturday Rock
Green Lookout Mtn
Cougar Rock
Little Lookout Mtn
Lookout Mtn
McKinley Ridge
Poison
Copper Creek
Star Cr
Miners Cr
Dole
Cedar Creek
Rock Creek
Spotted Deer Mtn
GPNF
Squaw Butte
Sturgeon Rock
Lookout
Silver Star Mtn
Little Baldy
Bluff Mtn
Pyramid Rock
Elkhorn Mtn
Larch Mtn
Last Chance Mine
Skamania Mine
Silver Cr
Prospector Cr
Washougal River
Stebbins Creek
Microwave Tower
CAMP BONNEVILLE
MILITARY RESERVATION
Spud Mtn
Creswell Heights
JEEP TR

The Washougal Mining District covers a portion of Clark and Skamania Counties. Gold was discovered in the 1890's on the East Fork of the Lewis River and the Washougal River. Placer gold in many streams indicated that rich deposits of lode gold should be in these mountains. This did not prove to be true. Some platinum was also found and low grade copper throughout the area.

Placers have been worked on the East Fork of the Lewis River, four to five miles east of Moulton. These placers also carry platinum. Bedrock is up to thirty feet deep in this area. See Yacolt Quadrangle for this section of Clark County.

Small amounts of placer gold may still be found on the East Fork of the Lewis River. From the placers previously mentioned east of Moulton, the Lewis River and Copper Creek should produce some gold for a recreational gold panner. The upper reaches of the Washougal River and its east flowing tributaries in this area should produce some gold. Yacolt and Lookout Mountain Quadrangles show this district.

Surveyor's Creek, a tributary of the Klikitat River, has produced some placer gold on the east slopes of Mt. Adams. See Mt. Adams East and Jungle Butte 7½' Quadrangles for this area of Yakima County.

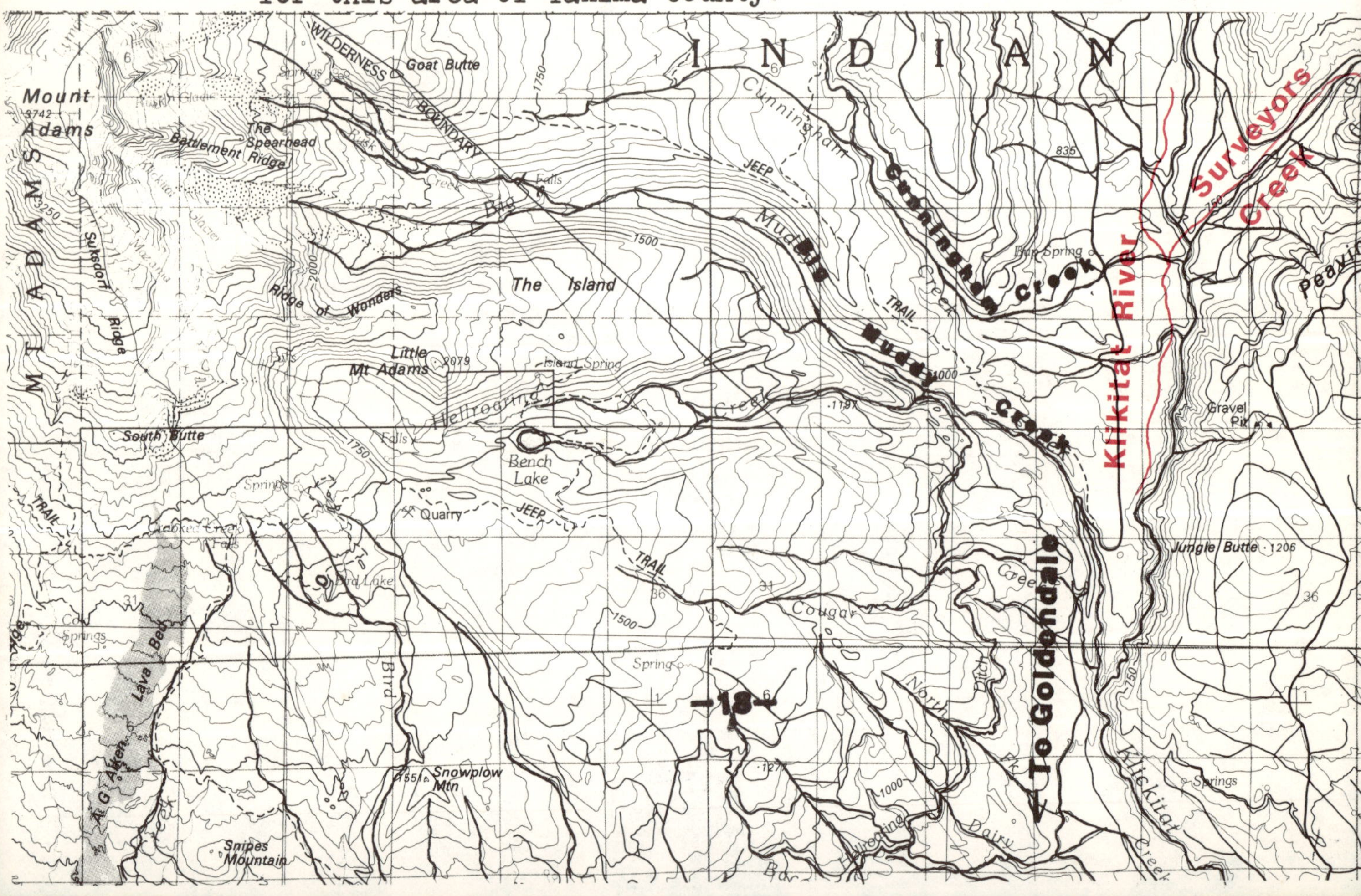

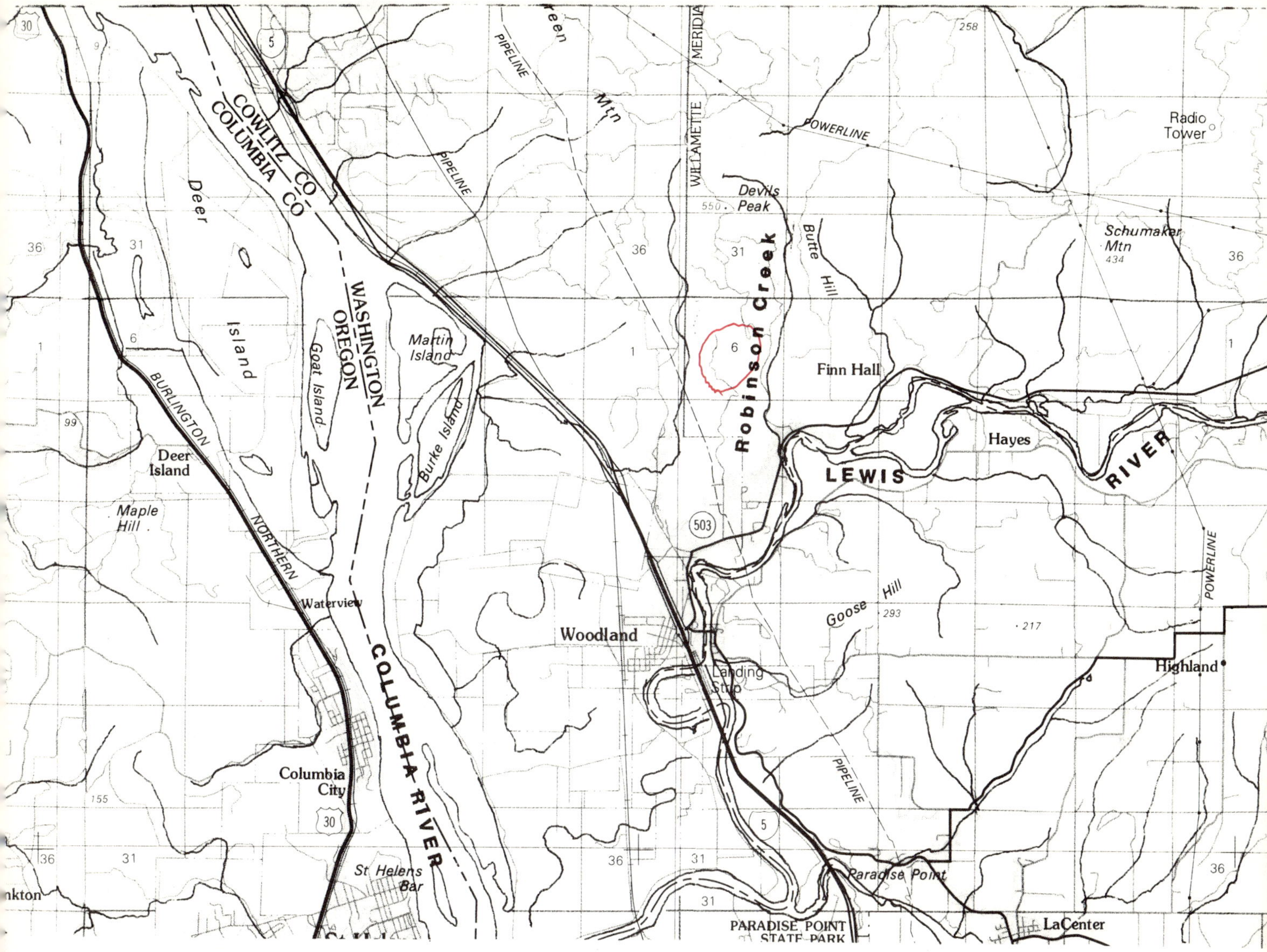

Traces of gold have been found on Green Mountain near Woodland. This is also shown on Yacolt Quadrangle.

A mineralized zone near Yale Bridge on Lewis River also carries low grade ore.

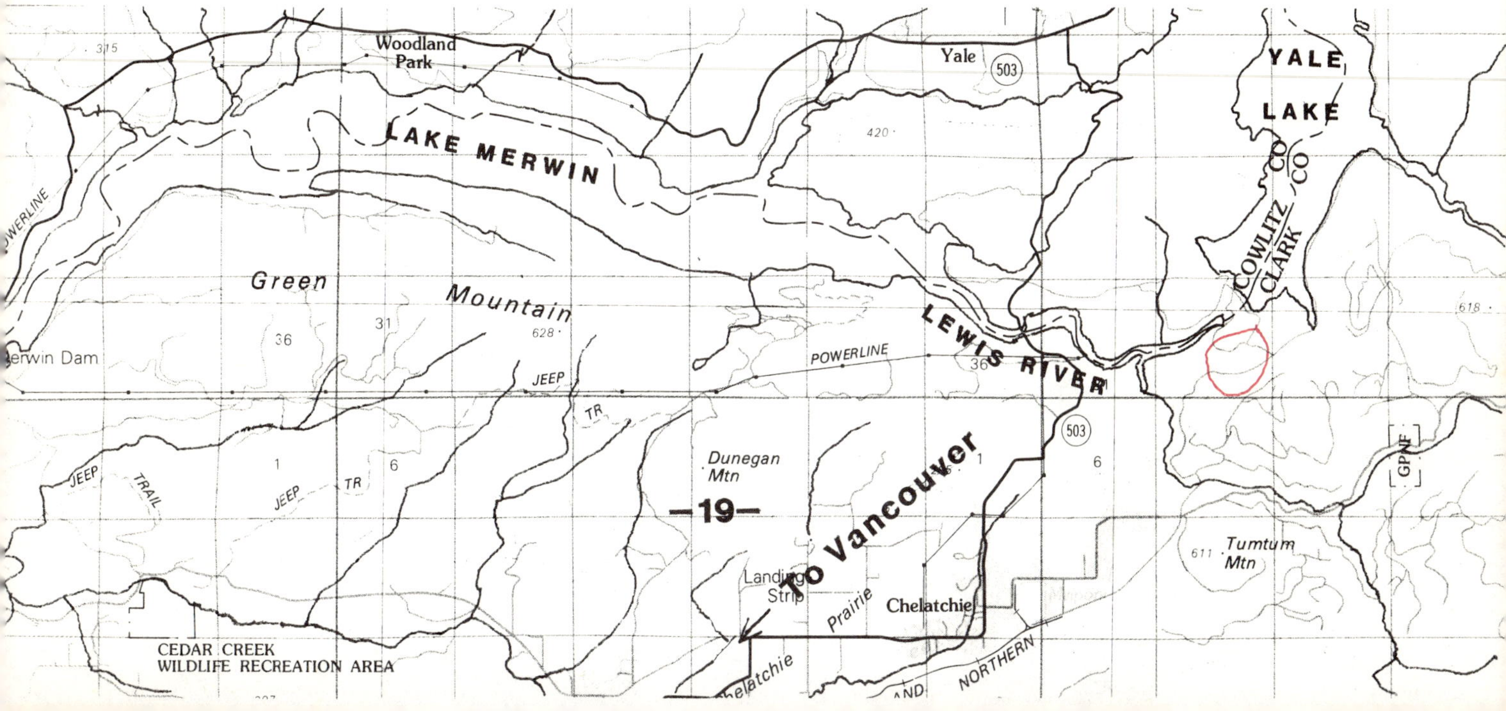

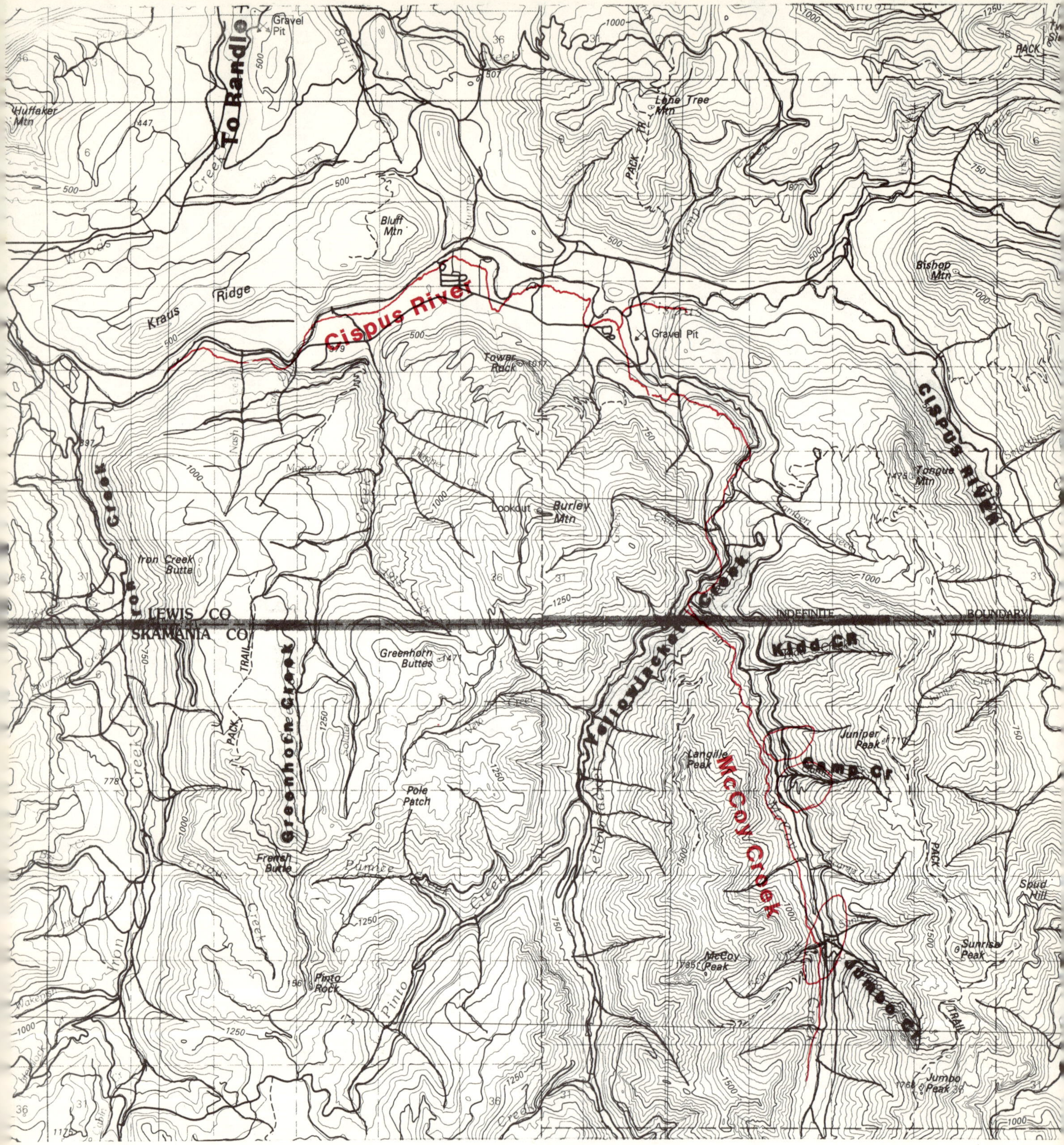

The McCoy Peak 7½' Quadrangle shows the most interesting prospect for gold in the Southern Cascades. Placer gold was discovered on the Cispus River and McCoy Creek in the early mining days, but production was minor. The depression miners revived the placers, but their production was also minor. Large boulders in McCoy Creek make placering difficult. Thin flakes of gold up to one-half inch diameter have been found here. Several placer claims have been staked along the Cispus River between Yellow Jacket and Juniper Creeks.

G I F F O R D P I N C H O T

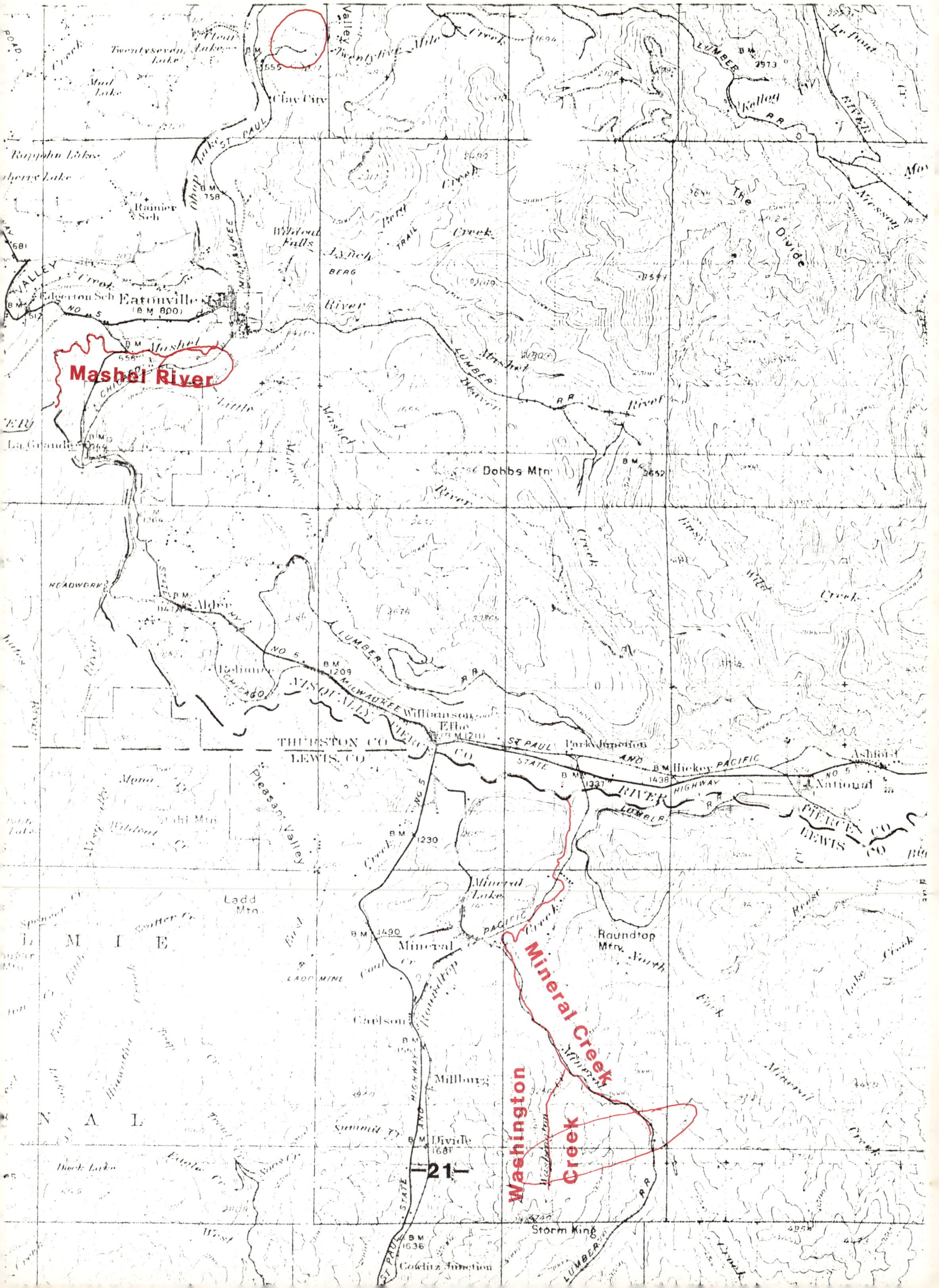
Mashel River
Mineral Creek
Washington Creek
Eatonville
Clay City
Twentyseven Lake
Mud Lake
Rainier Sch
Edgerton Sch
La Grande
Alder
Elbe
Dobbs Mtn
The Divide
Mineral
Mineral Lake
Roundtop Mtn
Ladd Mtn
Carlson
Millborg
Divide
Storm King
Cowlitz Junction
Park Junction
Hickey
Ashford
National
THURSTON CO
LEWIS CO
PIERCE CO
Pleasant Valley
NISQUALLY
STATE HIGHWAY
LUMBER R R
Mashel River

A belt of limestone south of Elbe carries values in gold and other minerals. This mineral zone is cut by Mineral Creek and Washington Creek. Streams in this area may produce some placer gold. See Mineral Quadrangle for this section of Lewis County.

The Mashel River has produced some placer gold near its mouth. Gold ore was produced from lodes three miles upstream. A wide vein of free milling quartz produced some low grade gold ore near Clay City. See Ohop Valley and Kapowsin 15' Quadrangles for this section of Pierce County.

Many streams flowing out of Mt. Rainier National Park show fine colors and traces of gold. The everlasting snow and glaciers may be concealing some interesting prospects. Any deposits inside the park are unworkable, but as long as these glaciers keep grinding rock the spring and summer floods will bring some of this gold outside the park.

In recent years placer gold has been found on the Carbon River near the Carbon River Park entrance. Mineralized areas along Cayada Creek, north of Mt. Rainier, may also feed some gold into the Carbon River. These are mainly copper prospects, but have some gold values. See Enumclaw Quadrangle for this area of Pierce County.

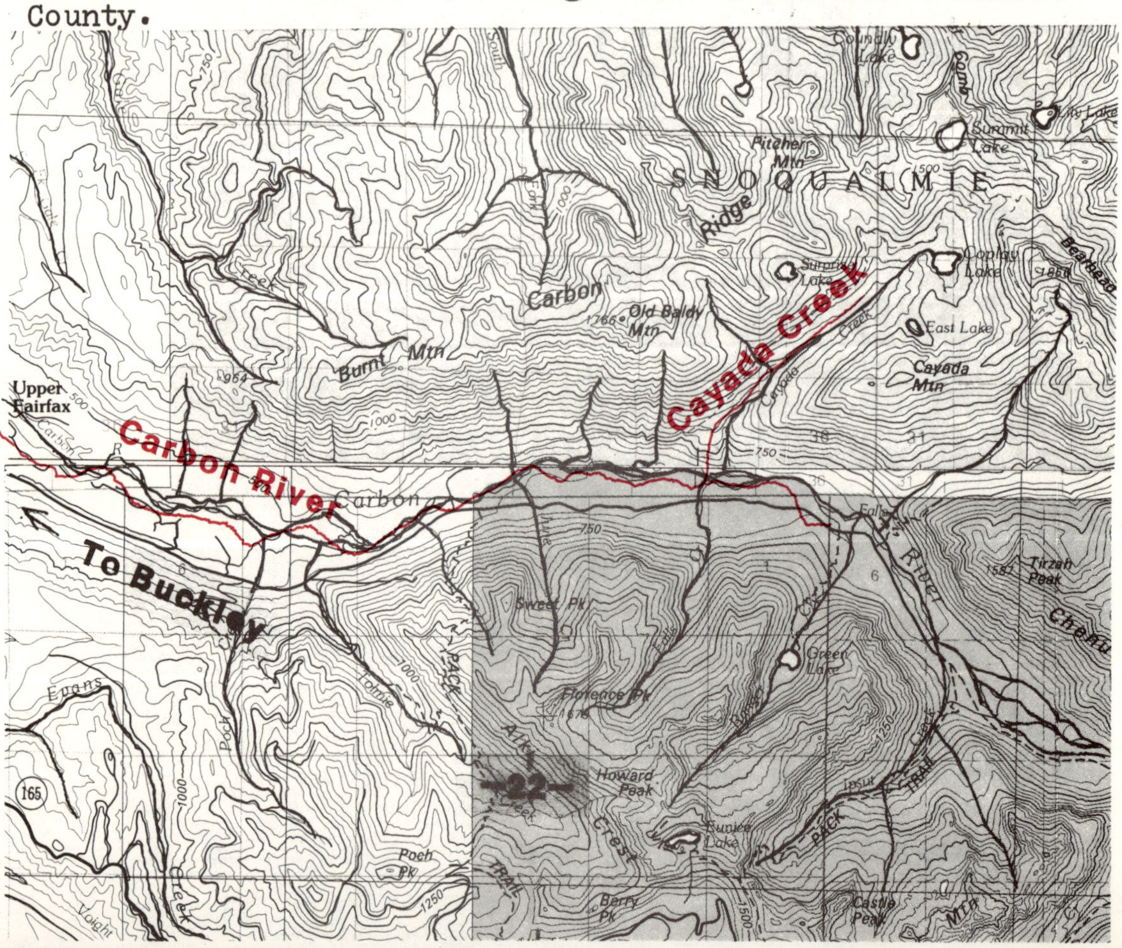

Gold has been found on Huckleberry Creek, a tributary of White River. This creek drains the north side of the park. Its headwaters are in the vicinity of Sunrise Lodge. Greenwater Quadrangle shows this area of Pierce County.

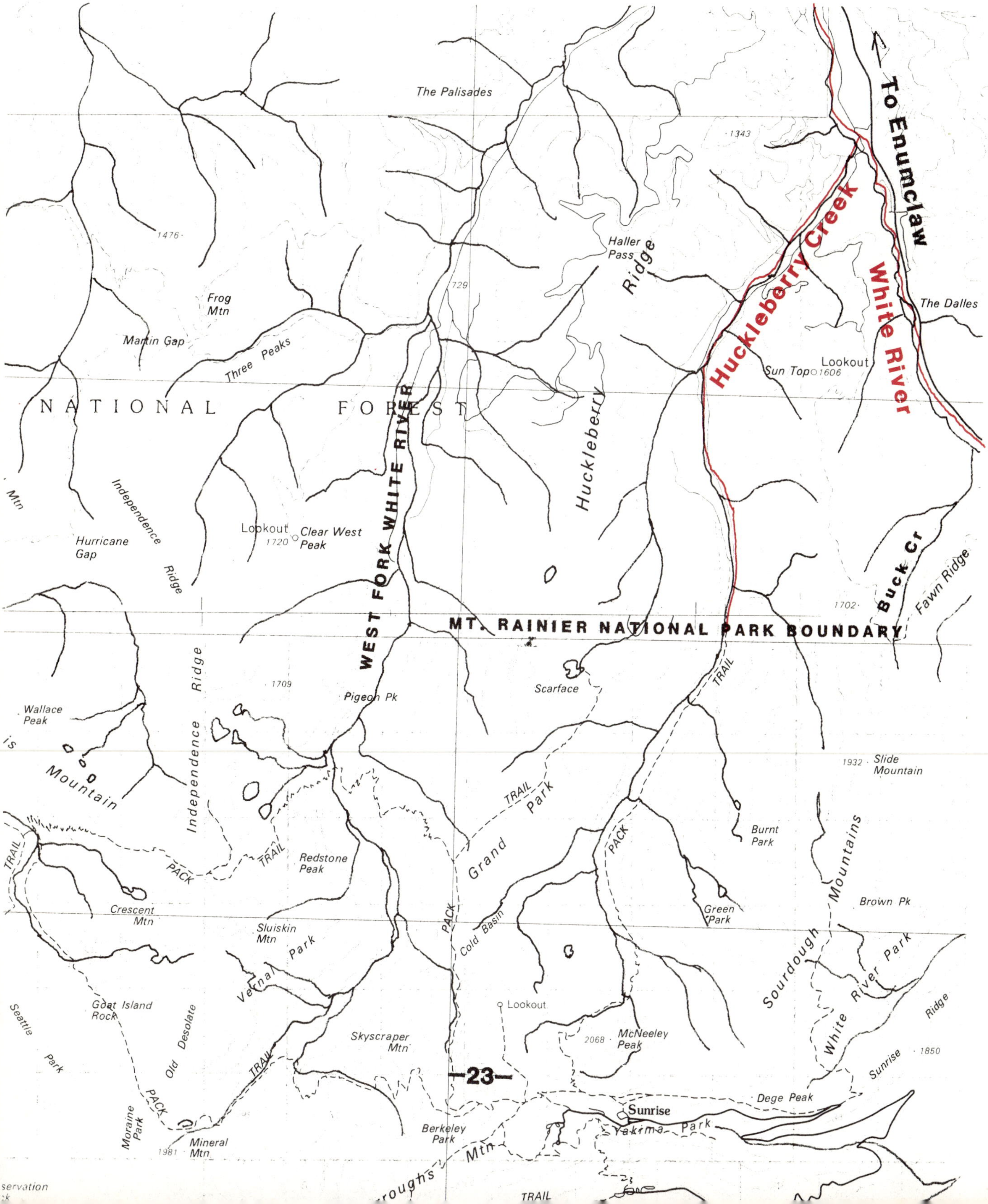

Gold has been found in many locations in King County. A piece of rough quartz showing wire gold was recently found by a fisherman on the Green River near the Kummer bridge. A gravel bar on the Green River at Palmer paid wages for a miner during the 1930's depression.

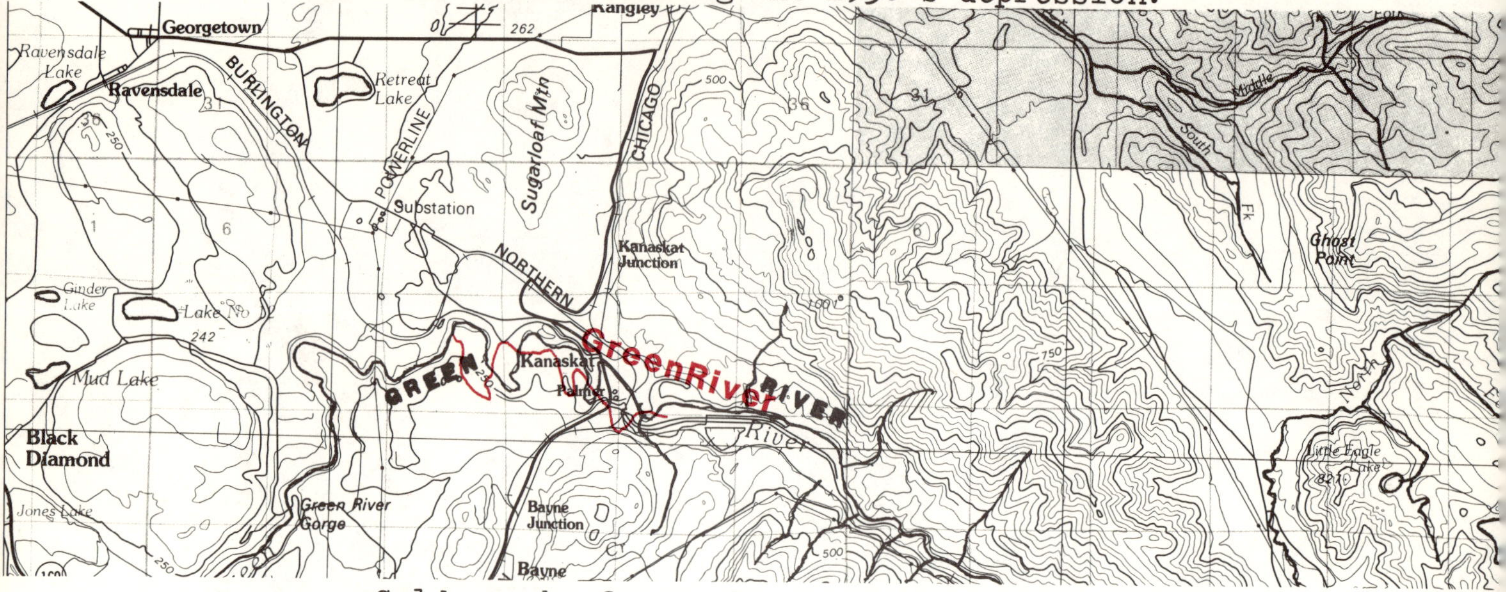

Gold may be found along the White River. This is usually very fine colors. The White River drains the northeast side of Mt. Rainier. This area is mineralized on both sides of the river with copper, silver, and gold. Before the National Park was established, the Storbo Mine was producing at the head of White River. Ore was packed out on mule back and the first road into the area followed the old pack trail.

Gold was found eight miles east of Enumclaw. Over three hundred feet of tunnel was dug and a stamp mill was built in 1890. Eagle Gorge and Enumclaw Quadrangles show these areas.

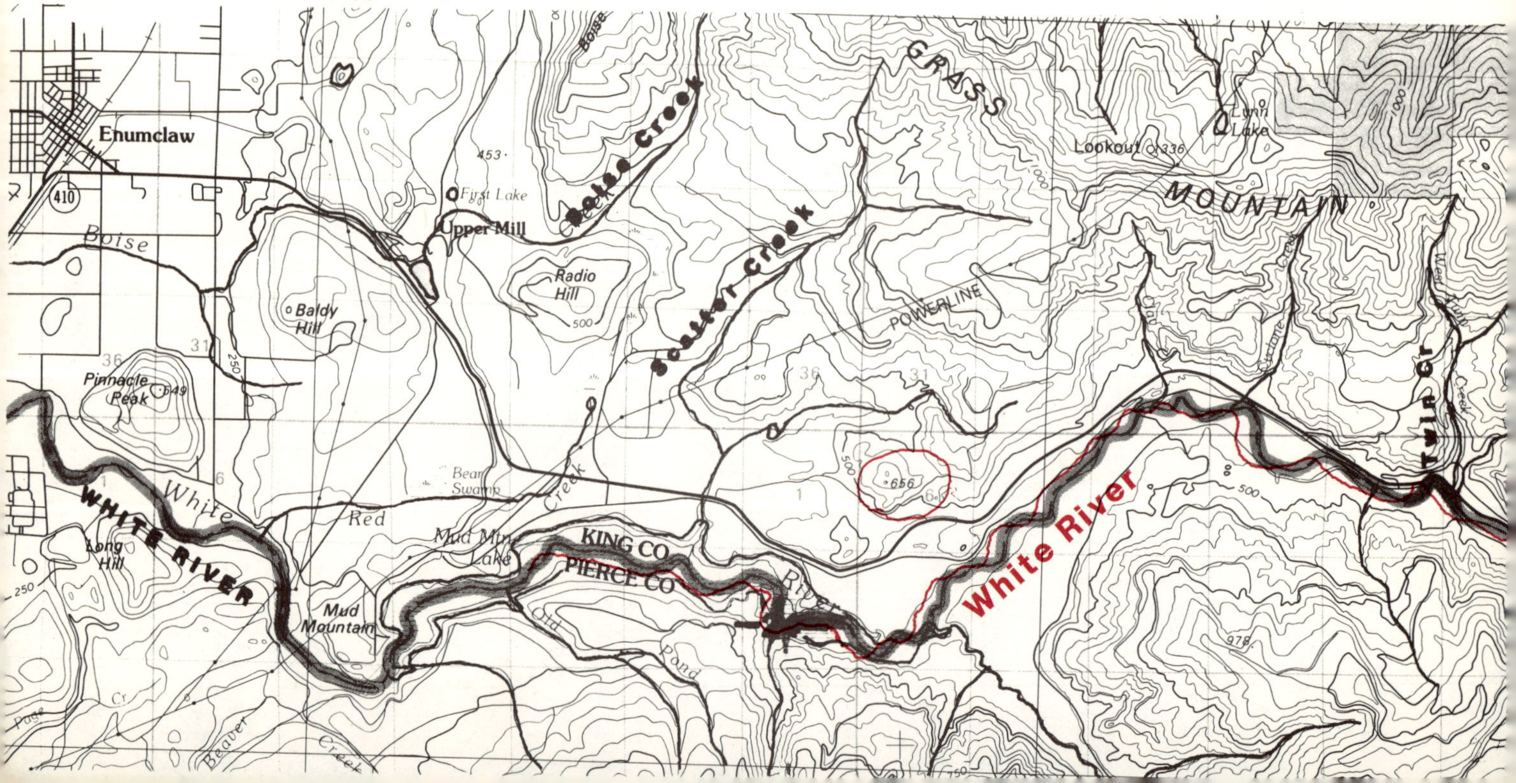

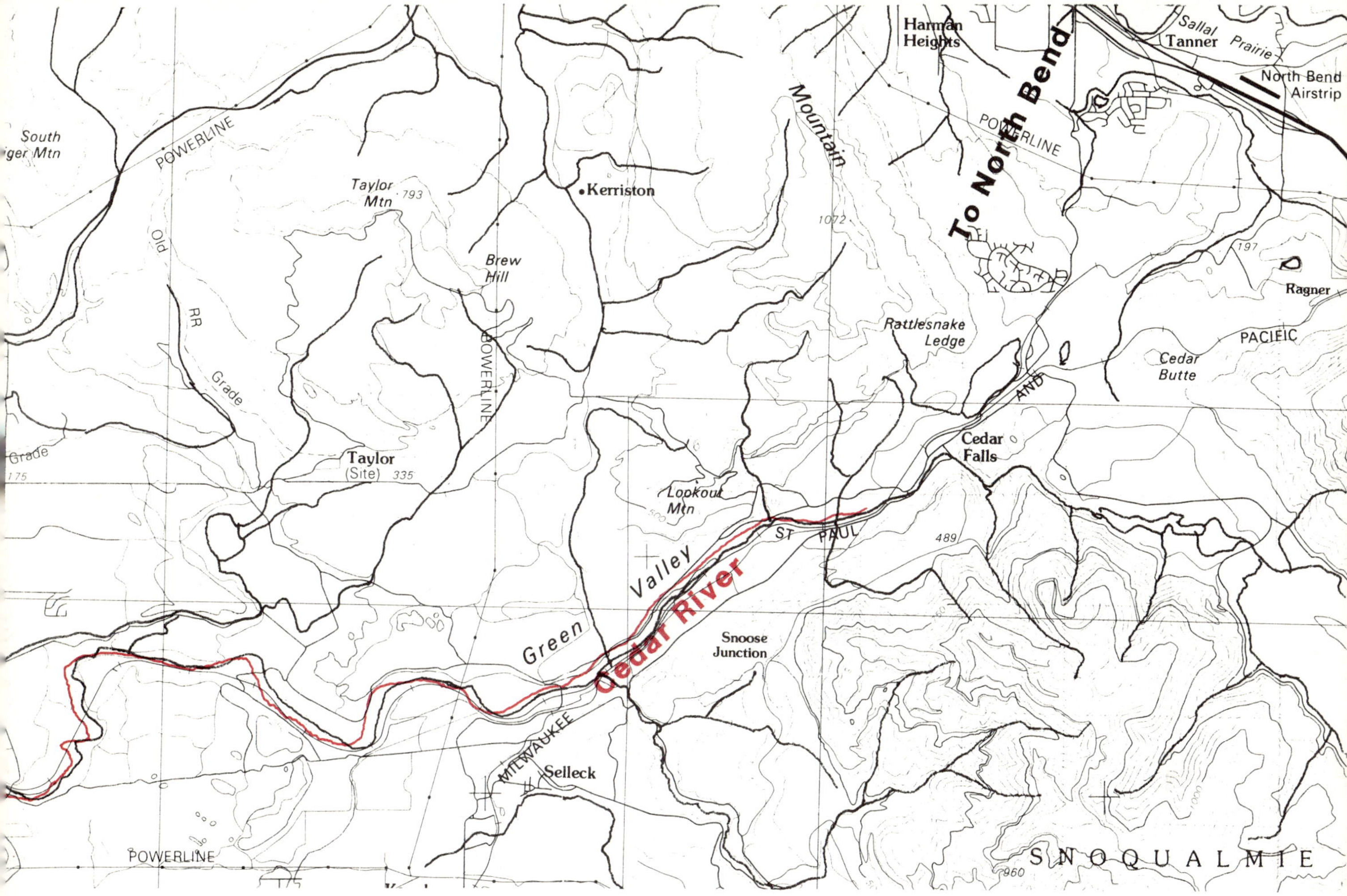

Some lode gold was found on the Cedar River and gold panners find an occasional color along with a lot of black sand.

Recreational gold panners find the Raging River to be productive. Flood waters seem to replenish the gold on these bars each year. Gold has been found the full length of this river, but the most productive area seems to be within a mile or two each direction from Preston. Fall City, Hobart and North Bend $7\frac{1}{2}$' Quadrangles show this river.

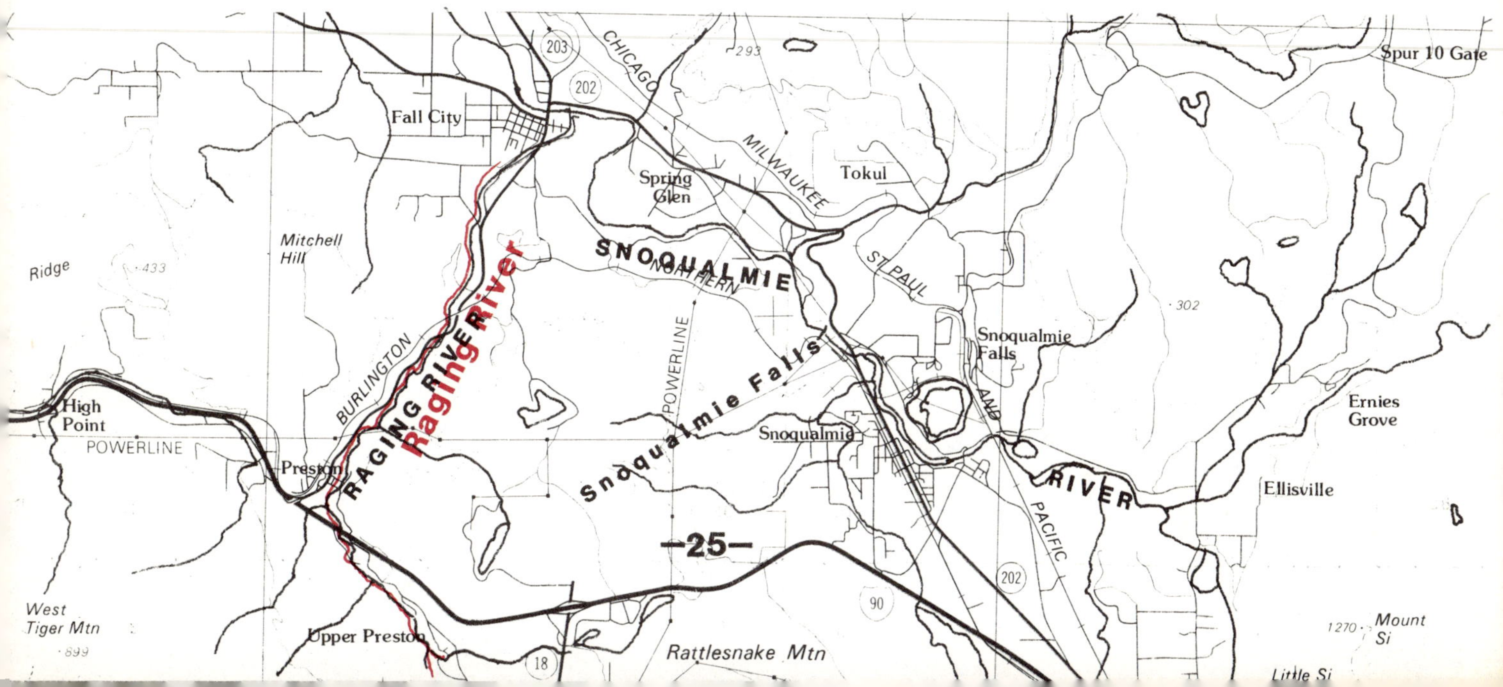

SUMMIT MINING DISTRICT

The winter snows and year round beauty of this area make it one of Washington State's most popular playgrounds. Chair lifts and ski trails now cover the site of feverish mining activity. Evidence of old mines and miner's cabins can still be seen on the slopes of Crystal Mountain. One old miner was asked if he really expected to find gold here. He said, "No, but I can't think of a prettier place to look for it."

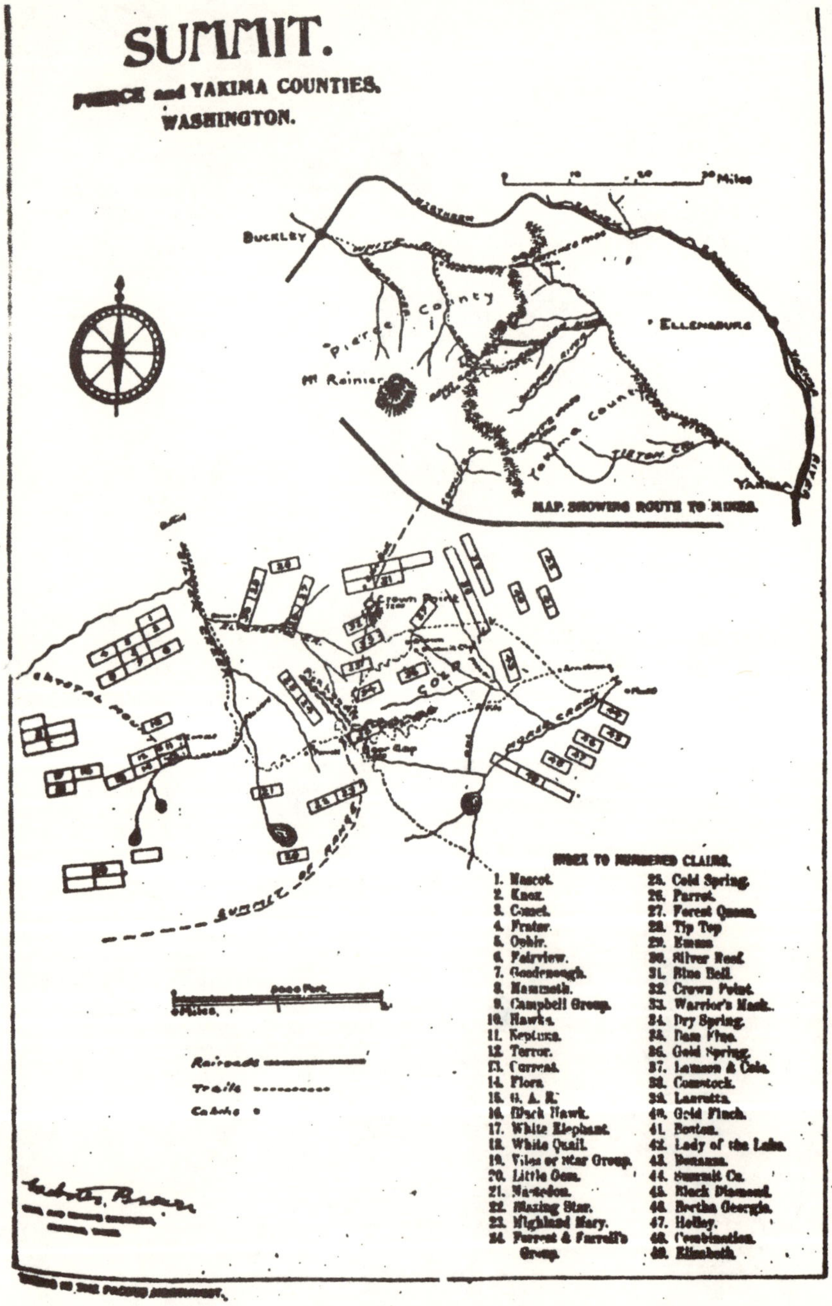

Silver Creek drains this basin and flows into White River. Recent mining activities on this creek was a top secret mid-winter dredging operation. Amount of production, if any, is unknown. This operation was brought to a halt when it was reported that someone had waded through four feet of snow and stolen all their equipment. Mt. Rainier and Bumping Lake Quadrangles show Crystal Mountain and Silver Creek in Pierce County.

The Summit Mining District crosses the high ridge into Yakima County. Morse Creek drains this area. A portion of Morse Creek was hydrauliced. Several large nuggets were recovered, but the cost of mining was greater that the values recovered.

Several lode mines showed high assay values. Some prospects are still being worked. The high expectations of the early miners are shown by their naming the mountain, Gold Hill. See Bumping Lake Quadrangle.

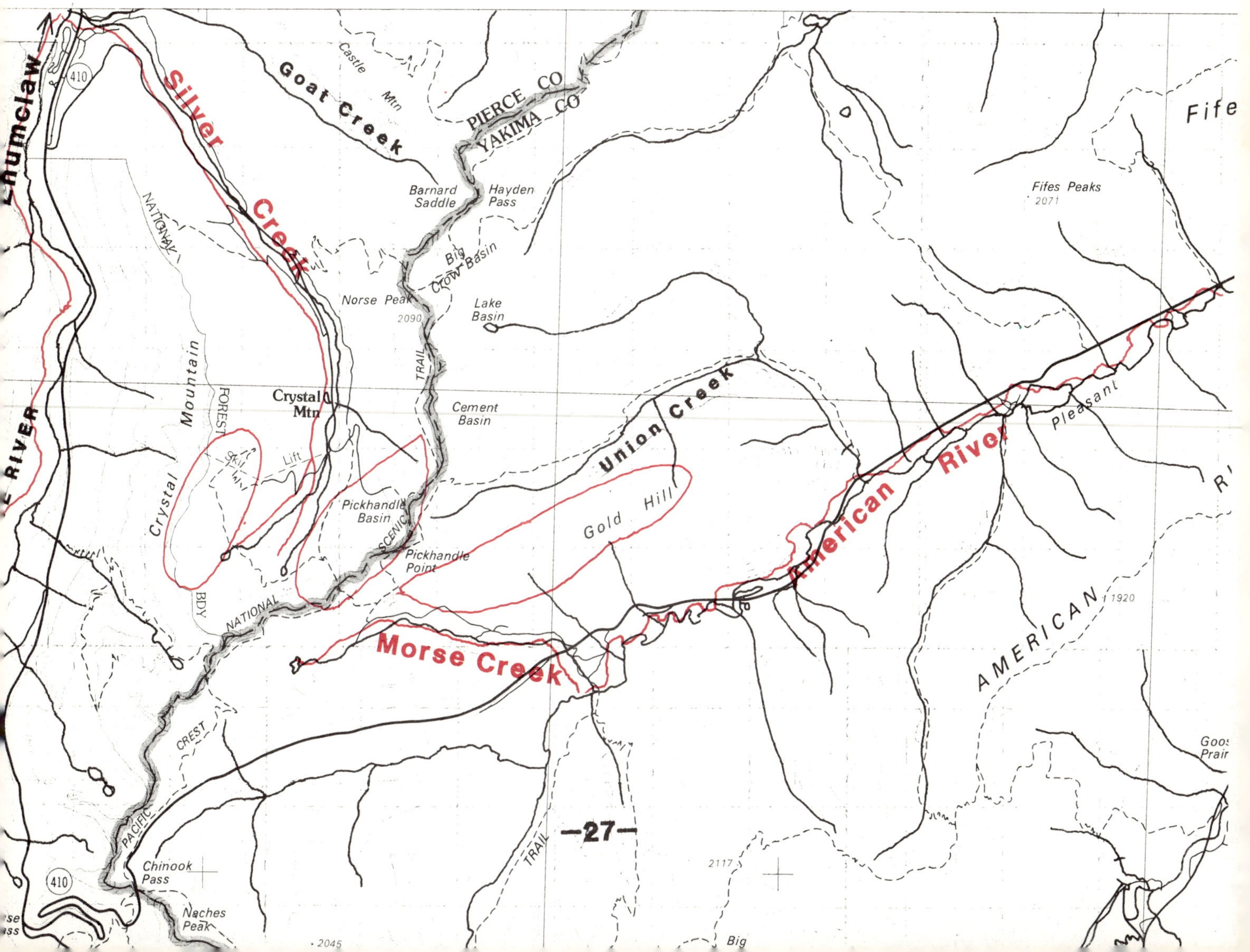

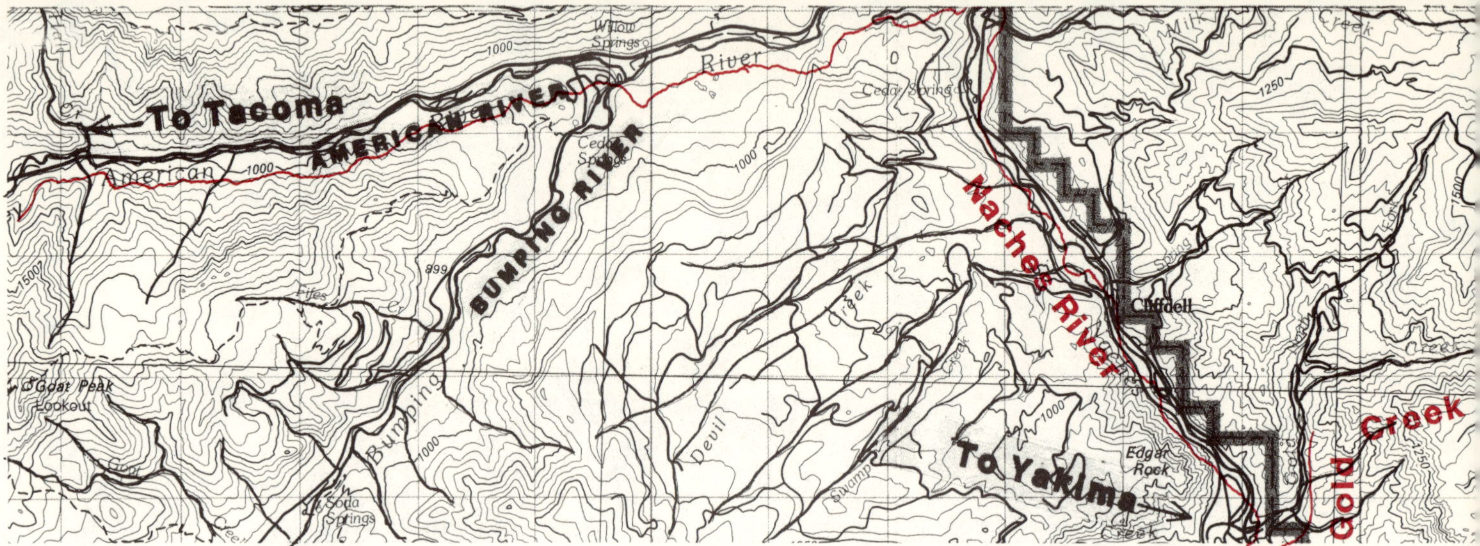

The Naches River drains a large area on the east slopes of the Cascade Mountains. Several place names along this river and its tributaries suggest mineral riches. Gold Run and Gold Creek are near Cliffdell. Some gold may have been found in stream gravels. Pyrites in the area may indicate minor finds of lode gold.

The headwaters of the American River are in the Summit Mining District. Here we find Miners Creek and Gold Hill. This area was described along with the Summit Mining District. Fine gold may be found in most streams flowing south to the river and on the bars along the American River. The Bumping Lake Quadrangle shows several mines and prospects in this part of Yakima County.

The Bumping Lake Quadrangle also shows the ghost town site of Copper City. This was primarily a copper prospect, but it had some gold values. Colors have been found in Copper Creek, Deep Creek, and Granite Creek.

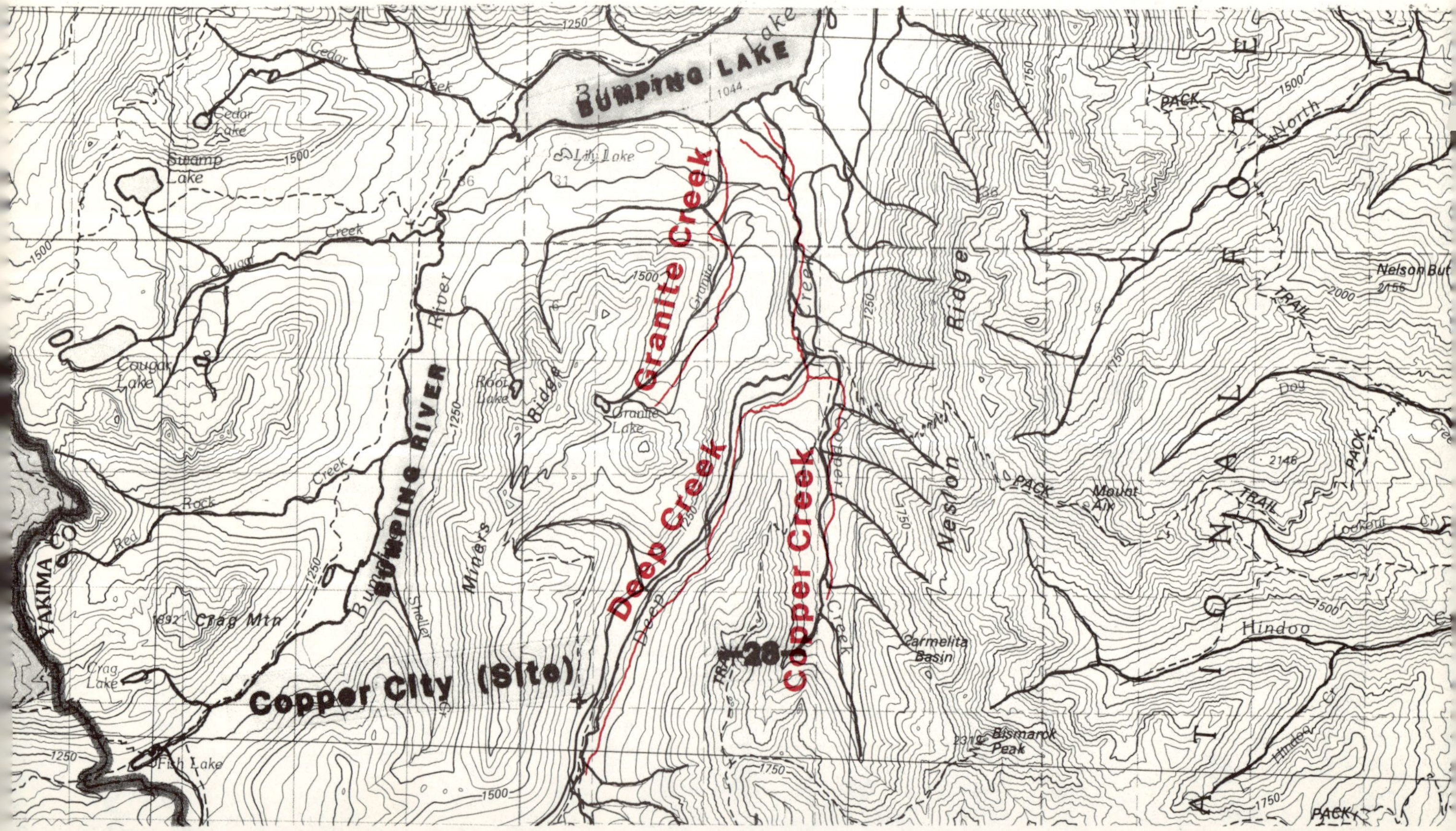

NORTH CENTRAL

The north central part of Washington State has produced more gold than all other parts of the State combined. The richest placers and the richest lodes are all found in this region. Areas producing placer gold in the 1860's are still producing. Areas producing lode gold in the 1870's are still producing.

The recreational gold panner can find gold in several places within an hours drive of Seattle or Everett. The sand bars along the South Fork of Snoqualmie River may show a few colors. Lode gold was found near the head of Hanson Creek above Bandera Airstrip. Lode mines were also worked on Denny Creek. Bandera and Snoqualmie Pass Quadrangles show this area.

East of Snoqualmie Pass, Gold Creek flows into Keechelus Lake. Near its head we find Silver Creek and Gold Lake. Several mines and prospects have been worked in this area.

Over the high ridge to the east, on Mineral Creek, ore was packed down a steep trail to the head of Lake Kachess, then barged down the lake to the old Snoqualmie Pass Wagon Road. This was a copper mine, but it also produced some gold. These mines are shown on the Snoqualmie Pass Quadrangle in Kittitas County.

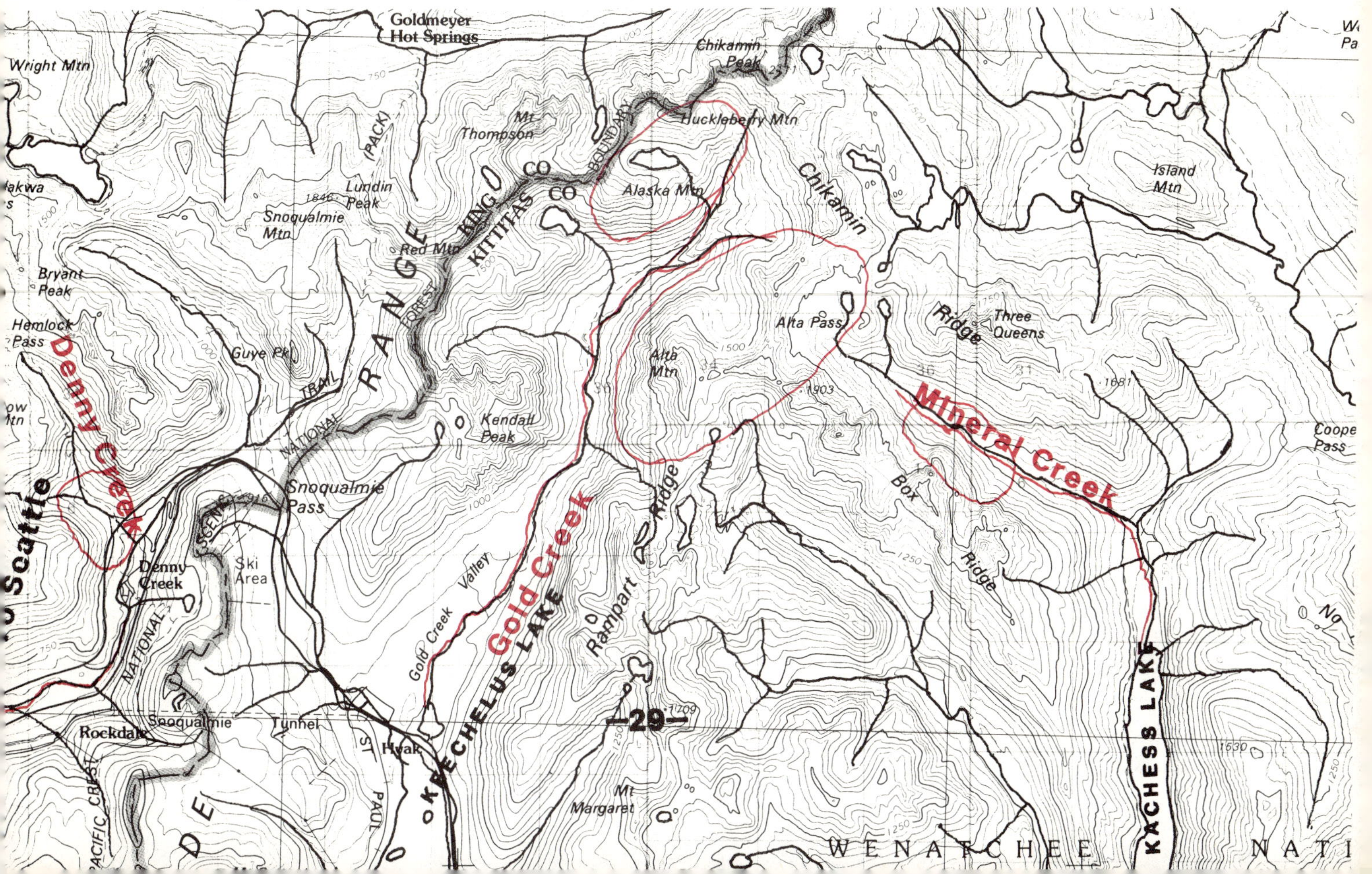

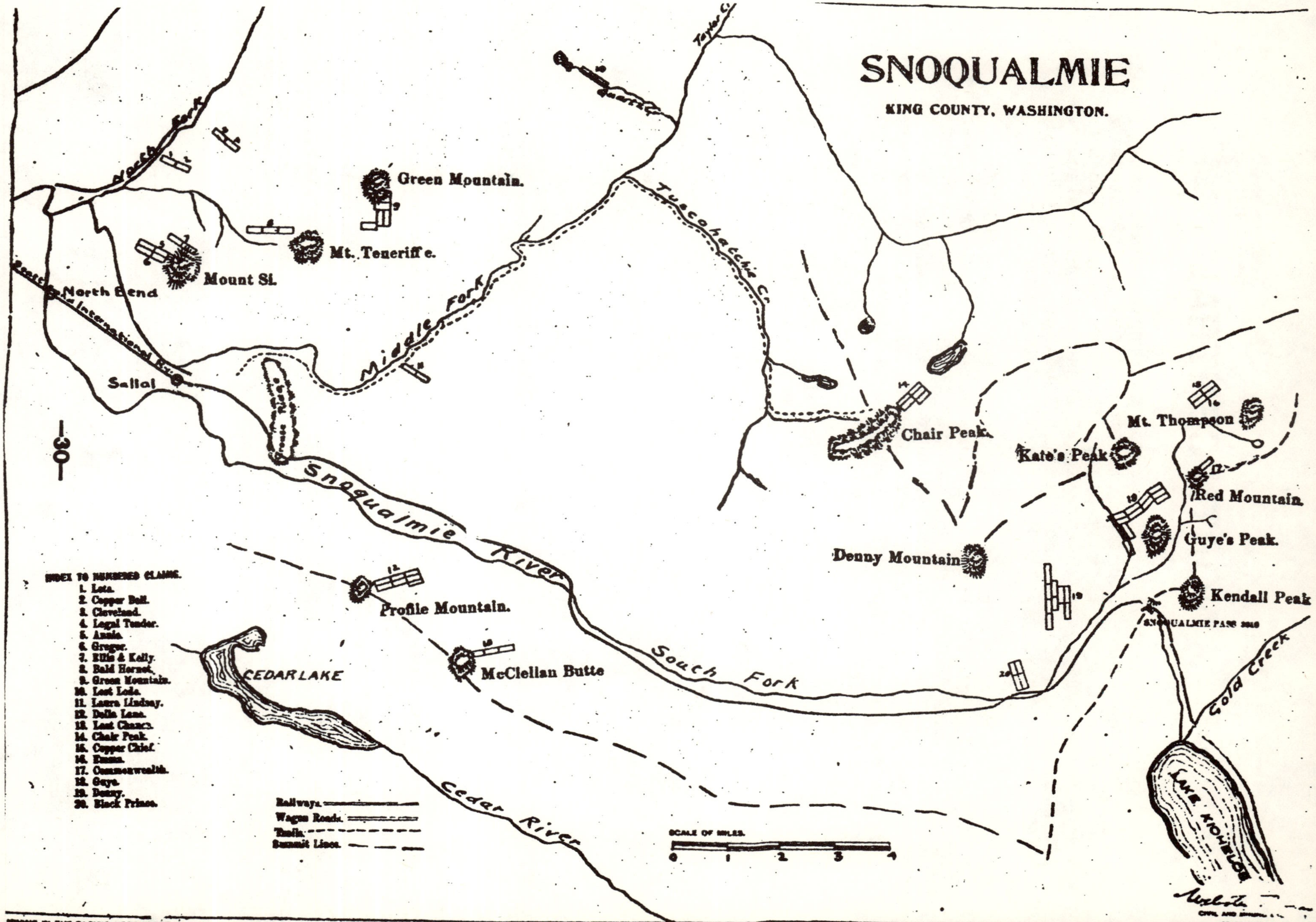
SNOQUALMIE
KING COUNTY, WASHINGTON.
Taylor C.
Quartz C.
North Fork
Green Mountain.
Mt. Teneriffe.
Mount Si.
North Bend
Sallal
Middle Fork
Tuscohatchie Cr.
Chair Peak.
Kate's Peak
Mt. Thompson
Red Mountain.
Guye's Peak.
Kendall Peak
Denny Mountain
SNOQUALMIE PASS
Snoqualmie River
South Fork
Gold Creek
Lake Kichelos
Profile Mountain.
McClellan Butte
CEDAR LAKE
Cedar River
INDEX TO NUMBERED CLAIMS.
1. Lota.
2. Copper Bell.
3. Cleveland.
4. Legal Tender.
5. Annie.
6. Gregor.
7. Ellis & Kelly.
8. Bald Hornet.
9. Green Mountain.
10. Lost Lode.
11. Laura Lindsay.
12. Della Lane.
13. Last Chance.
14. Chair Peak.
15. Copper Chief.
16. Emma.
17. Commonwealth.
18. Guye.
19. Denny.
20. Black Prince.
Railways.
Wagon Roads.
Trails.
Summit Lines.
SCALE OF MILES.
0 1 2 3 4

Gold has been mined along the Cle Elum River from its headwaters to its mouth. The Fortune Creek, Silver Creek, and Fish Lake area has produced considerable placer and lode gold. The Camp Creek drainage area had several lode mines. The Cle Elum River was placered for almost a mile near the mouth of Camp Creek. The Salmon La Sac area was placered extensively and had a large population of Chinese miners.

The upper end of Lake Cle Elum should produce gold for a dredge. The Cle Elum River is a wild river until it is slowed by the dam across the river. The spring run off should replenish this area with a new source of gold each year.

The large areas of gravel and sand near the mouth of the Cle Elum River have produced considerable gold. This large placer also extends down the Yakima River for several miles. Lake Kachess, Easton, and Cle Elum Quadrangles show this area of Kittitas County.

The Teanaway River has a few gold producing areas. Jungle Creek may produce a few colors or small nuggets. River bars for several miles upstream may show colors. Some good opportunities for dredging are along this river. The water is quite shallow in late summer and may be dredged without diving. See Mount Stuart Quadrangle.

The Yakima River has produced gold from the mouth of the Teanaway River to several miles below the mouth of Swauk Creek.

The Yakima River has the distinction of producing the first gold found in Washington State. Captain McClellan's railroad survey party found gold, but the discovery site is unknown. Gold has been found the full length of this river. The most productive areas are near the mouth of some of the rivers and creeks flowing into the Yakima.

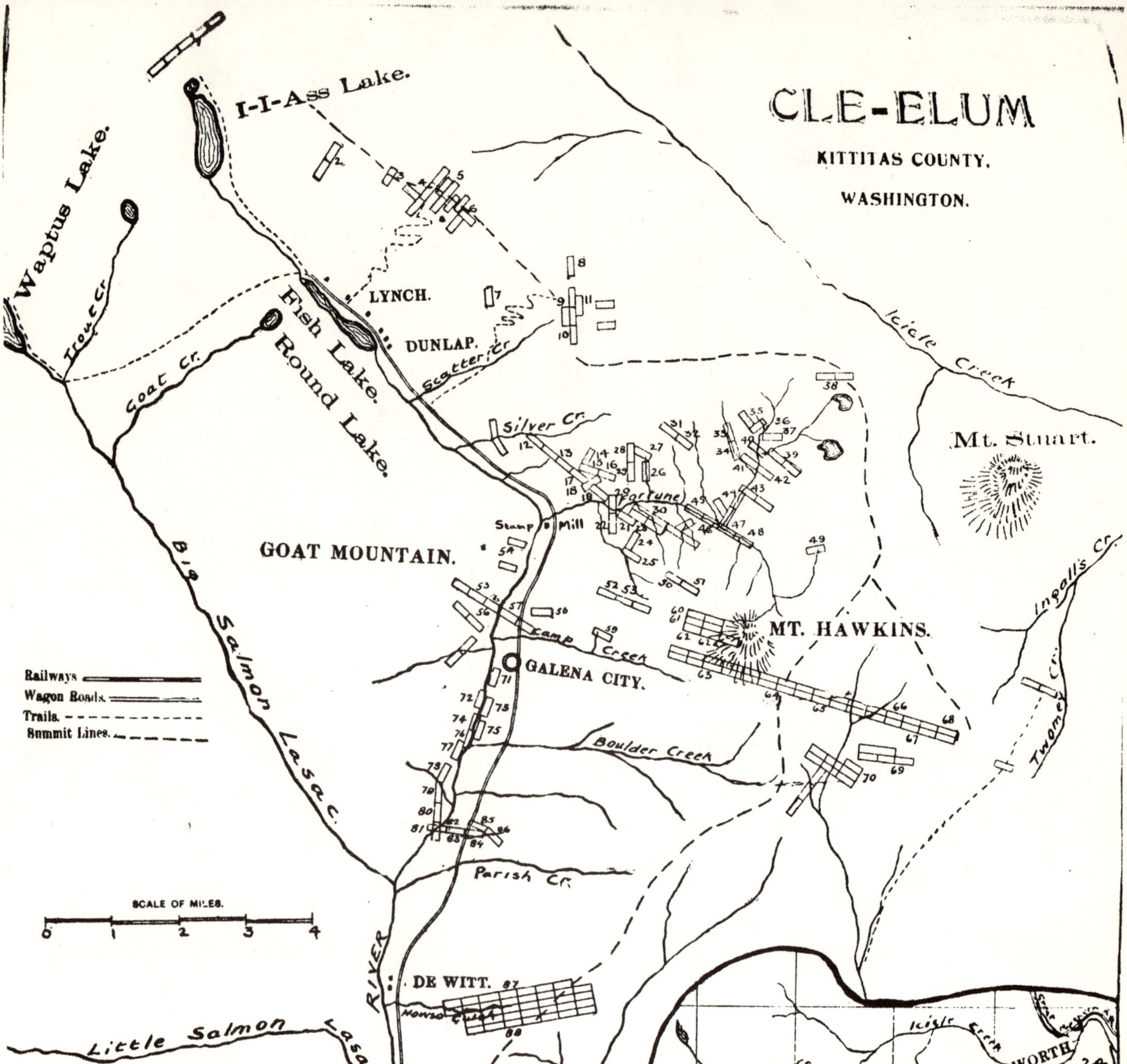
CLE-ELUM
KITTITAS COUNTY,
WASHINGTON.
I-I-Ass Lake.
Waptus Lake.
Trout Cr.
LYNCH.
DUNLAP.
Fish Lake.
Round Lake.
Goat Cr.
Scatter Cr.
Icicle Creek
Silver Cr.
Fortune
Mt. Stuart.
Stamp Mill
GOAT MOUNTAIN.
Big Salmon Lasac
Camp Creek
MT. HAWKINS.
GALENA CITY.
Ingalls Cr.
Twomey Cr.
Railways
Wagon Roads.
Trails.
Summit Lines.
Boulder Creek
Parish Cr.
SCALE OF MILES.
0 1 2 3 4
River
DE WITT.
Little Salmon Lasac
RED MOUNTAIN
N
PREACHER'S CAMP.
WELSH.
Cle-Elum
Teanaway River
Cle-Elum Lake.
NEWPORT.
DAVIS.
CIVIL AND MINING ENGINEER.
SEATTLE, WASH.
LEAVENWORTH
Mt. Stuart.
Ingalls Creek
Negro Creek
BLEWETT
Cle-Elum River
Cle-Elum Lake.
Teanaway River
RONALD.
ROSLYN
CLE-ELUM.
Northern Pacific Railway
Swauk Creek
MAP SHOWING ROUTE TO MINES.
R.14E 15 16
T.19N
24 22 21 20

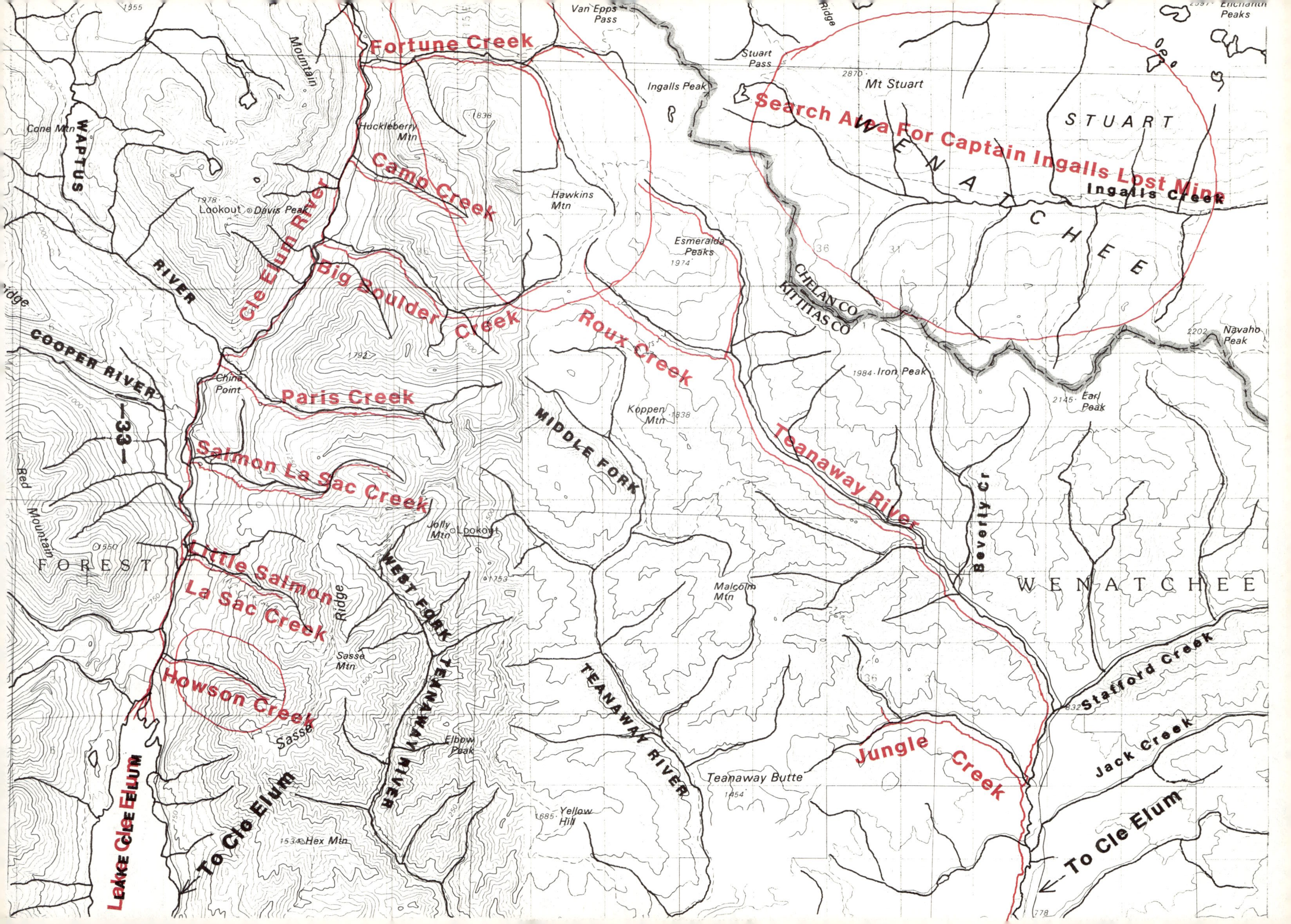
Search Area For Captain Ingalls Lost Mine
Fortune Creek
Camp Creek
Big Boulder Creek
Cle Elum River
Paris Creek
Salmon La Sac Creek
Little Salmon La Sac Creek
Howson Creek
Roux Creek
Teanaway River
Jungle Creek
Lake Cle Elum
LAKE CLE ELUM
To Cle Elum
To Cle Elum
Ingalls Creek
Stafford Creek
Jack Creek
Beverly Cr
STUART
WENATCHEE
WENATCHEE
FOREST
MIDDLE FORK
TEANAWAY RIVER
WEST FORK
TEANAWAY RIVER
COOPER RIVER
WAPTUS
RIVER
CHELAN CO
KITTITAS CO
Mt Stuart
Ingalls Peak
Stuart Pass
Van Epps Pass
Enchantment Peaks
Navaho Peak
Earl Peak
Iron Peak
Koppen Mtn
Esmeralda Peaks
Hawkins Mtn
Huckleberry Mtn
Lookout
Davis Peak
Cone Mtn
Mountain
China Point
Jolly Mtn
Lookout
Malcolm Mtn
Teanaway Butte
Yellow Hill
Elbow Peak
Sasse Mtn
Sasse Ridge
Sasse
Hex Mtn
Red Mountain
Ridge
—33—

LIBERTY
SWAUK MINING DISTRICT

The mining area around Liberty is an unusual and fascinating place. Gold can be found anywhere and in any formation. Gold ore has been found in sandstone, shale, basalt, quartz, calcite, and volcanic ash. The placers produce everything from gold dust to huge nuggets. This is one of the few places in the world that produces the beautiful crystalline wire gold.

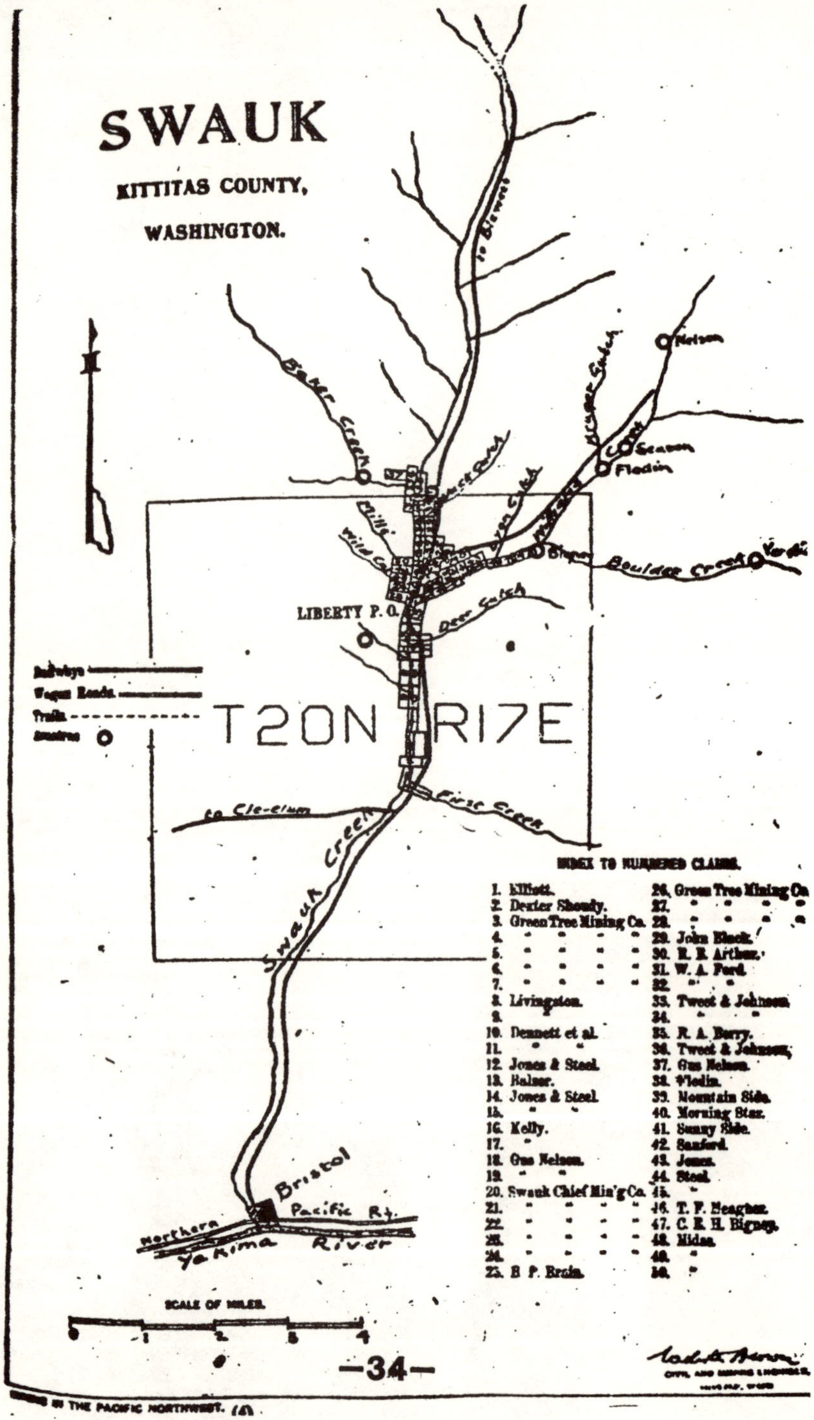

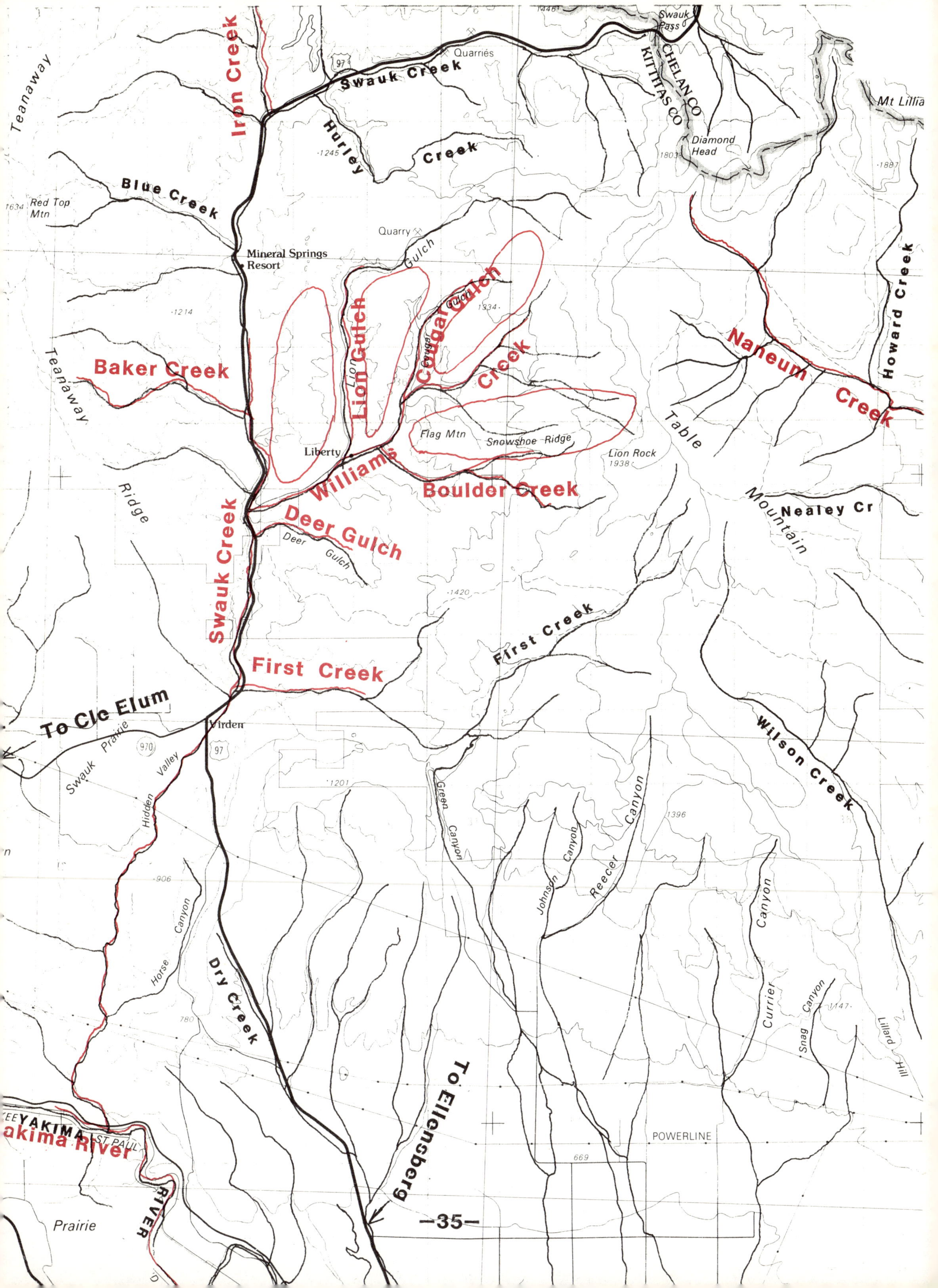
Swauk Pass
Swauk Creek
Quarries
97
CHELAN CO
KITTITAS CO
Mt Lillia
Diamond Head
1803
1887
1446
Teanaway
Iron Creek
Hurley
Creek
1245
Blue Creek
Red Top Mtn
1634
Quarry
Mineral Springs Resort
Gulch
Lion Gulch
Cougar Gulch
Cougar Gulch
1334
Creek
Naneum Creek
Howard Creek
1214
Baker Creek
Teanaway
Lion
Flag Mtn
Snowshoe Ridge
Lion Rock
1938
Table
Liberty
Williams
Boulder Creek
Ridge
Swauk Creek
Deer Gulch
Deer Gulch
Mountain
Nealey Cr
1420
First Creek
First Creek
To Cle Elum
Virden
970
97
Swauk Prairie
Valley
Hidden
1201
Green Canyon
Canyon
Johnson Canyon
Reecer Canyon
1396
Wilson Creek
906
Horse Canyon
Dry Creek
780
To Ellensberg
Currier Canyon
Snag Canyon
1147
Lillard Hill
YAKIMA
ST. PAUL
Yakima River
RIVER
POWERLINE
669
Prairie

You can expect to find gold in the most unexpected places. Some experts have stated that there is no geologic reason for gold to be here.

Quiet streams meander throughout this area. Most can be stepped over without getting your feet wet. Swauk Creek, Williams Creek, Boulder Creek, and Baker Creek have been the major producers of placer gold.

This area shows evidence of a violent geologic past. Sandstone, shale, and fossil tropical plants indicate that a vast lake or sea once covered this area. A wild river six hundred to eight hundred feet wide once flowed through here.

The bedrock of the ancient river is sandstone and shale. Basalt dikes have been thrust up in several places, blocking the river channel. A large amount of gold has been recovered from this old channel.

This old river is far up the mountainside and runs parallel to Williams Creek. In the vicinity of the intersection of the Liberty road with U.S. Highway 97, the bedrock is at the same elevation in both the old channel and Williams Creek. Bedrock is exposed under the Swauk Creek bridge at this point. A short distance upstream, bedrock is up to ninety feet deep.

At the west edge of the town of Liberty, bedrock is shallow again. After crossing Williams Creek, bedrock is again very deep.

Some sections of this old channel have been very rich. Others are barren of gold. Where bedrock was shale, the gold was trapped. On slick sandstone it passed on to form a rich pocket further downstream.

Deep shaft mines are necessary to work the old channel. A miner was very lucky if his shaft hit bedrock on shale. Many large nuggets have been found here. The size of many of these early finds was given in dollar value rather than weight. This was fine as long as the price of gold was fixed, but has little meaning today.

The largest nugget was found on Baker Creek. It had a value of $1004.00. Swauk gold has a high silver content. While the price of gold was fixed at $20.67 per ounce, Swauk gold was worth about $15.00 per ounce. Based on these figures this $1004.00 nugget weighed approximately 67 ounces. A nugget of this size today would be worth $28,000.00.

Large nuggets were common. Many valued at $100.00 to $200.00 were reported. One young miner's wife gave her baby a nugget to use for a substitute teething ring. The nugget was so large there was no danger of the baby swallowing it.

The ancient channel of this river has only been traced for a few miles. If a new unworked section of this river can be located, it may produce undreamed of wealth.

There are several theories about the origin of this coarse gold found in the old channel. Some think it has a local source. Others think that the hammered and rounded condition of these large nuggets indicate travel for many miles. If this is correct, where are other sections of the channel? Could this be a part of the Mammoth Channel that supplied the fabulous riches to the California 49'ers?

Waste dumps at many of the old mines have considerable gold. The best pay streak was in a few inches above bedrock. Any gold in the upper gravel was dumped with the waste. Several of these old dumps have been reworked profitably in recent years.

A part of Swauk Creek was dredged with a bucket dredge during the 1930's. This dredge used a trommel screen of a small size and large nuggets were dumped with the tailings. A few large nuggets have been found in these dredge tailings. A good metal detector might locate another.

Bedrock along Swauk Creek is shallow for much of its length. Bedrock is exposed under the Liberty Road bridge. This has been sampled by thousands of gold panners, but still shows colors. A tourist recently asked a local cafe owner to identify a pretty yellow rock she had found. It was a very nice gold nugget.

The Swauk was a good producer up to Baker Creek. The upper Swauk produced little. A small area on Iron Creek was placered. A lode prospect on Iron Creek is still being worked.

Baker Creek has produced a lot of gold. The benches north of the creek appear to be a part of the old channel. This pay area only runs about one mile above the junction with Swauk Creek.

Williams Creek has been very rich from its mouth to several miles above Liberty. Some miners are still working the old channel here. Deep shafts are required to reach bedrock. Several miners have tunneled under the road to reach the channel from the level of Williams Creek.

Lion Gulch has a few placers near its junction with Williams Creek. The ridge between Lion Gulch and Cougar Gulch had several lode mines. Some showed rich values.

Cougar Gulch had some of the more famous lode mines. Remains of several stamp mills and arrastras can still be found. Thousands of feet of hardrock tunnels have been dug here. No timbering was required in many of these mines. Many prospects are still active.

Flag Mountain and Snowshoe Ridge have produced a large amount of the famous Liberty wire gold. Ollie Jordin struck his legendary "wash tub" full of gold on Flag Mountain. His brother, Clarence also made fabulous strikes on Flag Mountain and Snowshoe Ridge.

Boulder Creek is the site of the most ambitious mining operation in the area. Buldozers and dragline cranes are being used to feed a high trommel washer. This plant has been worked on an intermittent basis for several years. Sand and gravel has been sold from this plant to pay expenses and any gold recovered is cream off the top.

The Liberty area has the potential to produce a lot of gold. Modern mining methods could be profitable working the old river channel. A good metal detector might locate a pocket of the beautiful wire gold. The prospects are good. As old Bull Nelson Said, "There is a lot of gold at Liberty, but there is an awful lot of gravel mixed up with it".

Gold has been found on Naneum Creek and other streams on Table Mountain, east of Liberty. This is usually in very thin flakes.

The Liberty and Thorp 15' Quadrangles show this area of Kittitas County.

Liberty Arrastra

BLEWETT MINING DISTRICT

Gold may have been discovered on Peshastin Creek as early as 1853. Historians do not agree on the discovery date, but by the mid 1870's this area was a bee hive of activity. Placer gold was found to be very rich on Peshastin Creek and many of its tributaries. Parts of this narrow canyon are almost completely filled by U.S. Highway 97 at this time, but gold can still be found here.

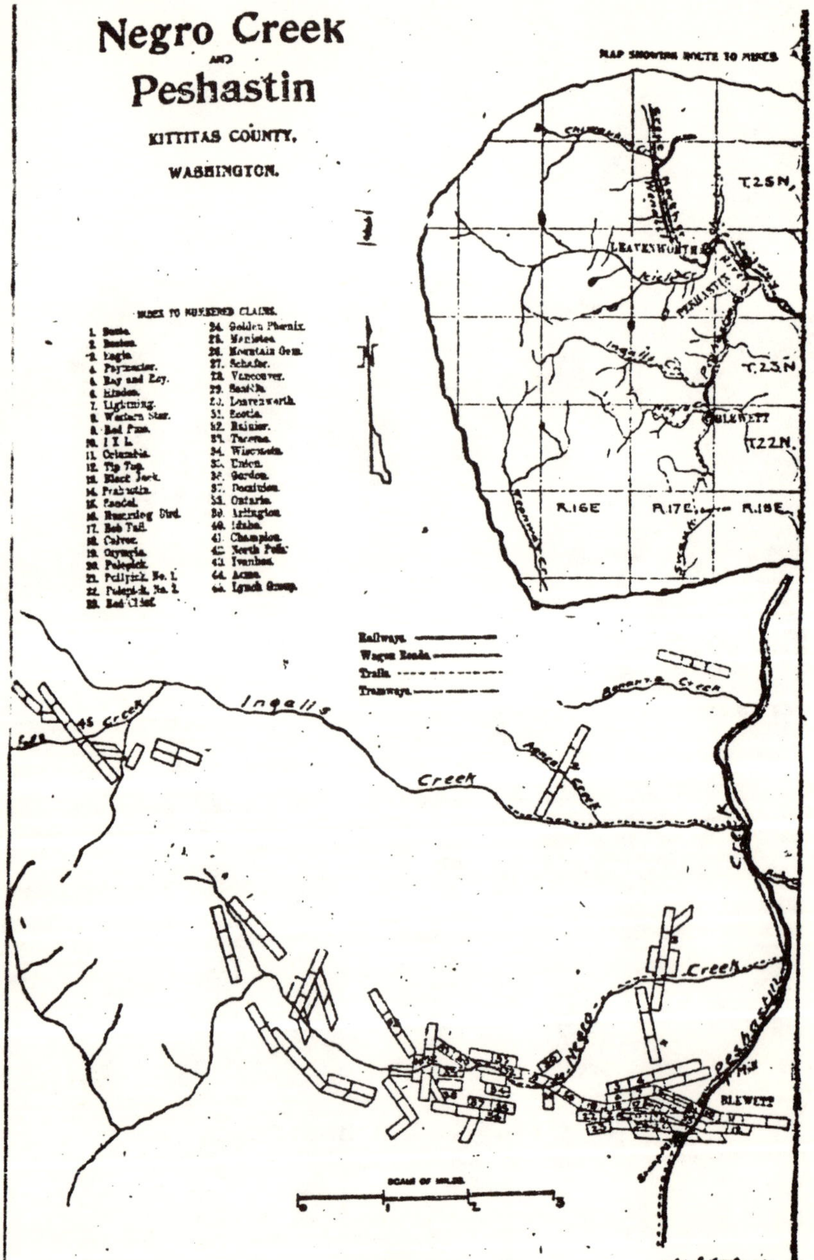

Many lode mines have been very rich in this district. Quartz and calcite veins up to eight feet wide are found throughout the area. Several stamp mills were built to process the rich ore. The remains of one mill can still be seen at the old townsite of Blewett. Many arrastras were also built. A well preserved one is now a historic landmark.

Lode gold was discovered in Culver Gulch in 1878. The first stamp mill in Washington Territory was built to process this ore. Culver Gulch is slightly over one mile in length. It has a difference of elevation of 1600 feet from Peshastin Creek to its head. Thousands of feet of tunnels have been dug in this short distance. Total amount of production is unknown, but it is believed to have been considerable.

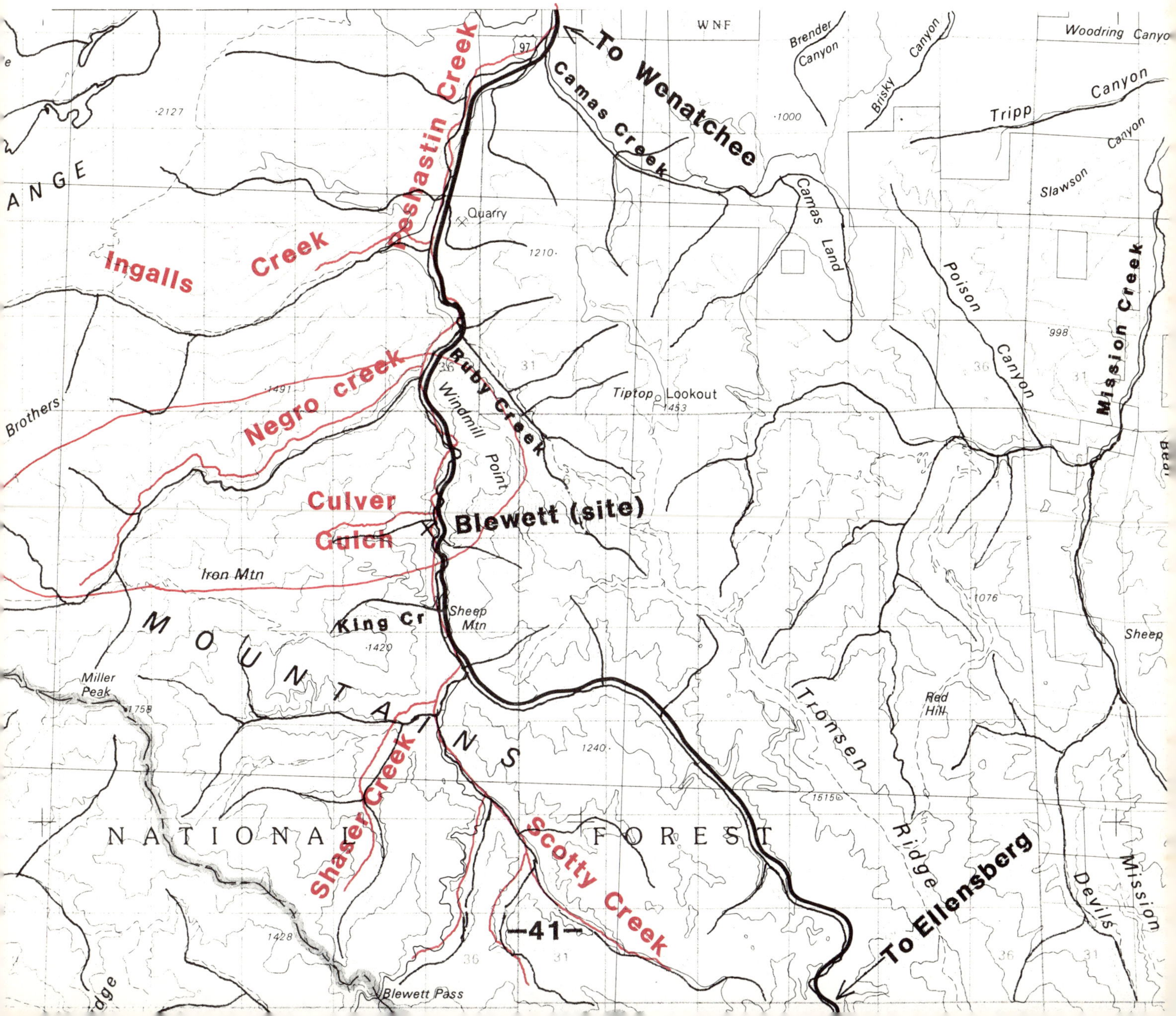

Gold may be found the full length of Peshastin Creek and in the Wenatchee River below its mouth. Coarse gold has been found near the headwaters. Bedrock is shallow in many places and good concentrations can be found in the bedrock cracks.

Scotty Creek will show good returns for your efforts. This stream produces small flake gold and fine colors. It has a wide valley and an unworked section might be located away from the present stream.

Peshastin Creek, below the mouth of Scotty Creek, carries an increasing amount of magnetite. This area has been under claim for its vast deposits of iron ore. Black sand can be a problem in sluice riffles. Particles of magnetite in the stream bed range from dust size to huge boulders.

Shaser Creek, from its junction with Peshastin Creek to its forks, is a consistant producer of thin flakes of gold. Bedrock is exposed for much of this distance. Bedrock cracks or any obstruction that causes gravel to collect will usually produce a few flakes of gold. Gold is continuously being replenished on this stream and each year will have a new supply of gold. The upper reaches of the South Fork Shaser Creek is also a good producer.

Peshastin Creek, downstream from Shaser Creek, has been heavily worked by dredge repeatedly over the last few years. Easy access and shallow bedrock makes this area very popular with the weekend prospectors doing a little recreational panning. They are usually rewarded with a few colors.

Lode mine tunnels may be seen on both sides of Peshastin Creek from the junction of the old Blewett Pass highway for several miles downstream. Some of these are still active on an intermittent basis. The modern day miner has the same dream as the old timer of hitting a rich pocket.

Rich placers have been worked near the mouth of Negro Creek. These placers extend for several miles upstream. Lode mines have been worked all along the ridge near the head of Culver Gulch and on upstream for several miles.

The mouth of Ingall's Creek and Peshastin Creek in the vicinity has been placered. Ingall's Creek was named for an Army Captain who was exploring in this area in 1872. Ingall's Creek drains a large area of the Mt. Stuart Range and somewhere in these mountains Captain Ingall's Lost Gold Mine awaits discovery.

The Blewett Mining District of Chelan County is shown on Liberty Quadrangle.

GOLD! GOLD! GOLD!
FEVER STRIKES WENATCHEE

This headline in the Seattle Post-Intelligencer greeted readers on Saturday morning, April 2, 1983. No, it was not a belated April Fool's Day joke. Gold has been re-discovered in the Wenatchee area and the rush is on. One thousand three hundred new mining claims were filed during the month of March and thousands more are expected to be filed before summer is over.

This area, south of Wenatchee, has a history of gold production dating back to the early 1890's. A large area is known to have low grade gold ore. Many prospect shafts, tunnels, and drifts have followed the elusive quartz veins for thousands of feet through the sandstone in this area.

The discovery of this body of gold ore is described by Hodges in "Mining in the Pacific Northwest". The fine particles of this gold and primative milling methods made recovery difficult. Early assays showed values up to $16.00 per ton, but very little mining was done at this time.

This deposit was extensively explored during the 1930's. In 1949 the difficulties of working this ore was overcome. Over the next eighteen years almost $15,000,000.00 in gold and silver ore was sold to the Tacoma smelter. The property has been more or less inactive since 1967.

Exploration and core drilling at depths far greater than any previous working has discovered several valuable deposits. Core drills have struck pockets assaying almost three ounces per ton. When the present owners of this property revealed their findings at their stockholders meeting, the rush was on to stake new claims all over the area.

The possibility of striking gold on any of these thousands of new claims has about the same odds as being the big winner in the State Lottery. The gold may be down there, but the necessity of digging hard rock shafts several hundred feet deep will elimate most would-be miners. It will require a ton of money to recover the first ounce of gold.

This strike is in a geologic formation known as Chumstick Sandstone. This formation is similar to the Swauk Sandstone and was considered as being the same until recent studies have proved a different age and composition between the two.

Bulletin 75, Geology of the Wenatchee and Monitor Quadrangles, by Randall L. Gresens, shows the area geology and maps of the Chumstick formation. This new publication can be a valuable aid to prospectors. Order from Washington State Department of Natural Resources, Division of Geology and Earth Resources, Olympia, WA 98504. Price $5.00 plus $1.00 postage.

Monitor, Wenatchee and Wenatchee Heights 7½' Quadrangles show this area of Chelan County.

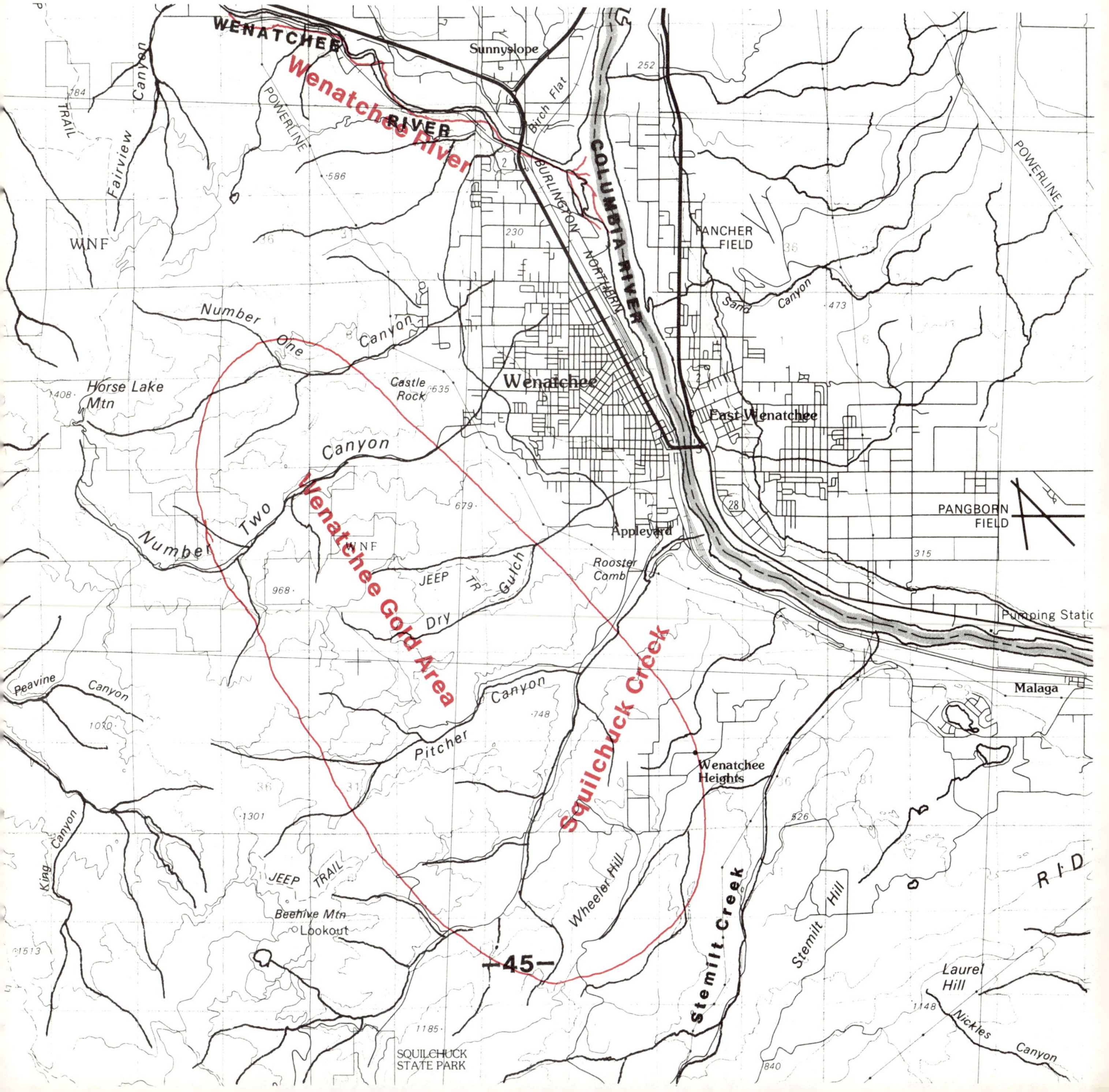

The Skykomish River and its tributaries have many good placer locations. Many of these are within easy driving range of the cities of Everett and Seattle. Some of these placers have been worked from the earliest mining days and are still producing gold.

The Sultan River will produce colors from its mouth to its source. Good placer locations have been found for a mile or so below the mouth of the Sultan River in the Skykomish River. The Sultan Canyon and Horseshoe Bend areas have produced considerable coarse gold. See Monroe and Index Quadrangle.

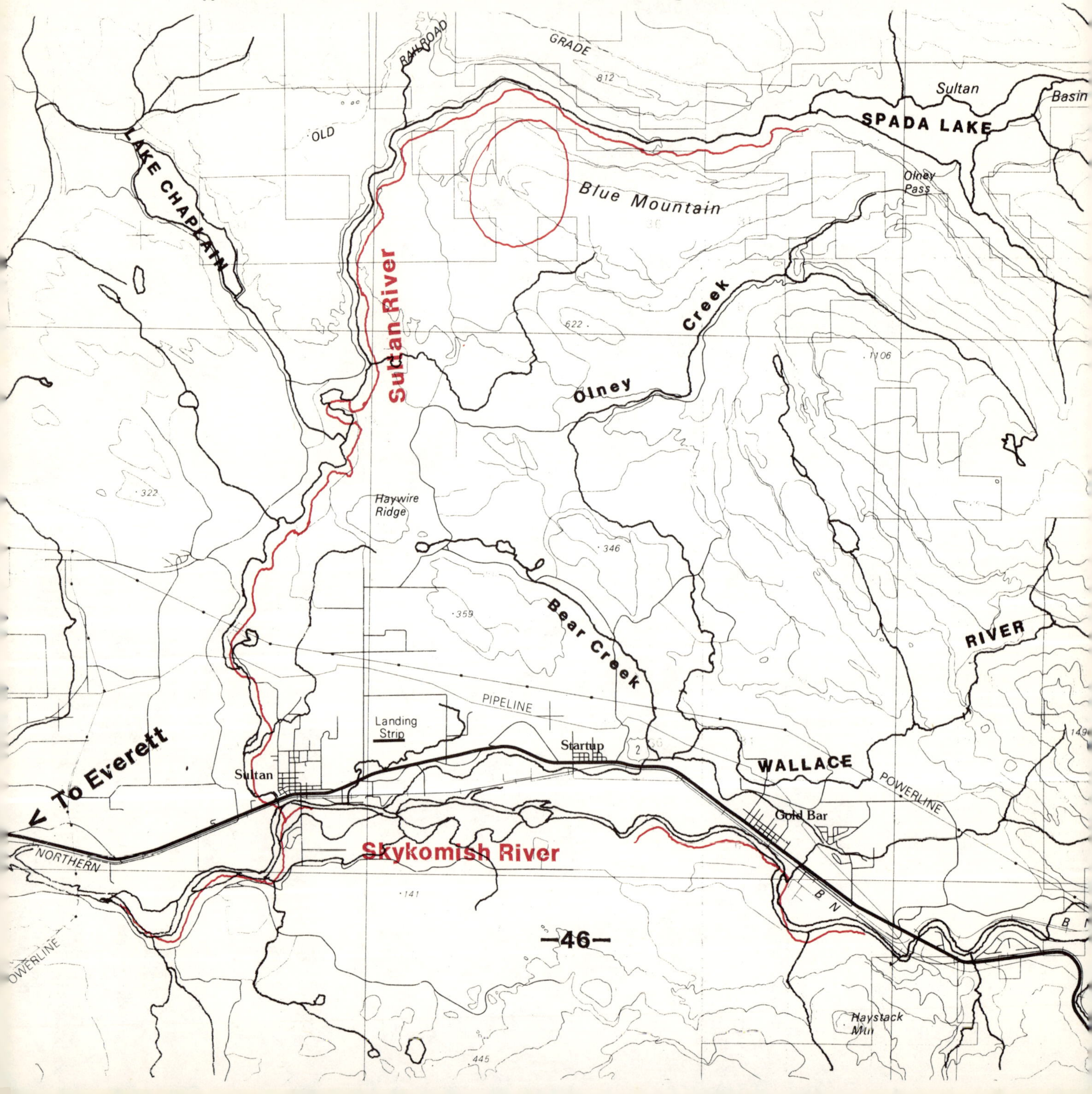

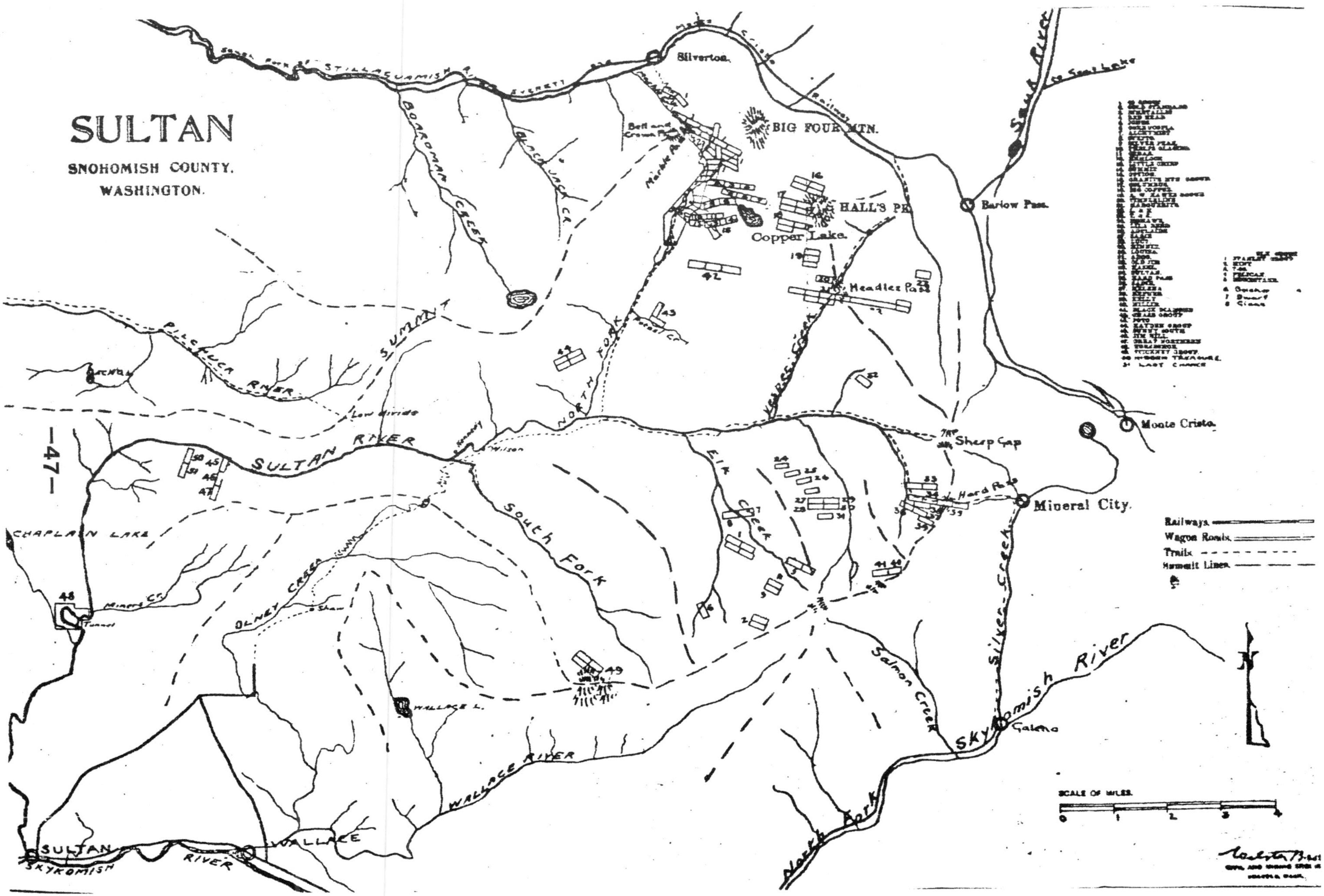
SULTAN
SNOHOMISH COUNTY,
WASHINGTON.
Silverton
BIG FOUR MTN.
HALL'S PK
Copper Lake
Headlee Pass
Barlow Pass
Monte Cristo
Sheep Gap
Hard Pass
Mineral City
Galena
SULTAN RIVER
South Fork
NORTH FORK
Elk Creek
Salmon Creek
Silver Creek
Skykomish River
WALLACE RIVER
WALLACE L.
WALLACE
SULTAN
SKYKOMISH RIVER
Sauk River
STILLAGUAMISH R.
BOARDMAN CREEK
BLACK JACK CR.
PILCHUCK RIVER
SUMMIT
Low Divide
CHAPLAIN LAKE
Miners Cr.
Tunnel
OLNEY CREEK
Railways
Wagon Roads
Trails
Hemmeth Lines
SCALE OF MILES
0 1 2 3 4
N

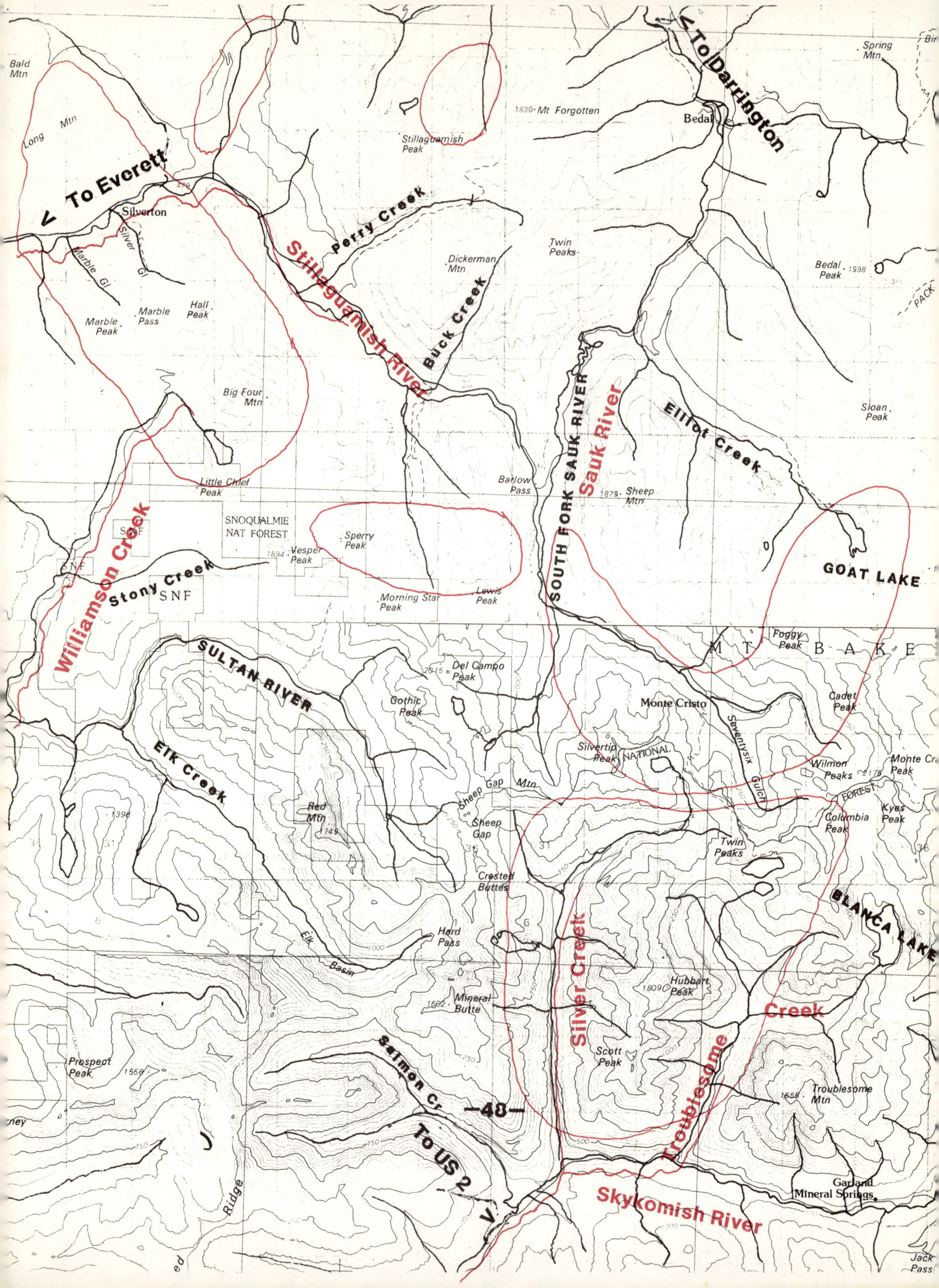

To Everett
To Darrington
Bedal
Silverton
Perry Creek
Stillaguamish River
Buck Creek
South Fork Sauk River
Sauk River
Elliot Creek
Goat Lake
Williamson Creek
Stony Creek
Sultan River
Elk Creek
Silver Creek
Troublesome Creek
Blanca Lake
Skykomish River
Salmon Cr
To US 2
48
Bald Mtn
Long Mtn
Mt Forgotten
Stillaguamish Peak
Spring Mtn
Twin Peaks
Dickerman Mtn
Bedal Peak
Marble Peak
Marble Pass
Hall Peak
Big Four Mtn
Sloan Peak
Little Chief Peak
Barlow Pass
Sheep Mtn
Snoqualmie Nat Forest
Sperry Peak
Vesper Peak
Morning Star Peak
Lewis Peak
Foggy Peak
Del Campo Peak
Gothic Peak
Monte Cristo
Cadet Peak
Silvertip Peak
Seventysix Gulch
Wilmon Peaks
Monte Cristo Peak
Columbia Peak
Kyes Peak
Red Mtn
Sheep Gap Mtn
Sheep Gap
Twin Peaks
Crested Buttes
Hard Pass
Elk Basin
Mineral Butte
Hubbart Peak
Scott Peak
Troublesome Mtn
Prospect Peak
Garland Mineral Springs
Jack Pass

The upper Sultan River drains a large mineral rich area. There are several lode mines along Williamson Creek. These lodes have produced considerable gold and other minerals. See Index and Silverton Quadrangles for this part of Snohomish County.

The town of Gold Bar was named for the gravel bars found on the Skykomish River. These placers extend for about three miles above and below the town. Production has been limited, but you might find a few colors. See Index Quadrangle.

The area around the town of Index has several lode mines. Some gold has been found, but mainly as a minor value in copper mines.

The North Fork of the Skykomish River has good placers in the vicinity of Galena. Howard Creek has produced placer gold near its mouth. The Silver Creek tributary of the North Fork has placers and lodes from its mouth to its headwaters. Most streams flowing into Silver Creek will show colors. This area is shown on Monte Cristo 7½' Quadrangle.

Monte Cristo. Even the name sounds rich. Monte Cristo was rich, but had a short life. Discovered in 1889, these mines produced $7,000,000.00 over the next fourteen years until the ore played out. A mountain of ore was in sight when Joseph Pearsall scaled the ridge at the head of Silver Creek and looked into the basin at the head of the Sauk River.

A railroad was built into the area and a network of tramways funneled ore to be shipped to the smelter at Everett. Hundreds of claims were filed on the Monte Cristo and Goat Lake ore bodies. Unfortunately the complex nature of these ores made recovery difficult. There was very little, if any, free milling ore in this area so placer mining was never a factor here.

This area is shown on Bedal, Sloan Peak, Monte Cristo and Blanca Lake 7½' Quadrangles.

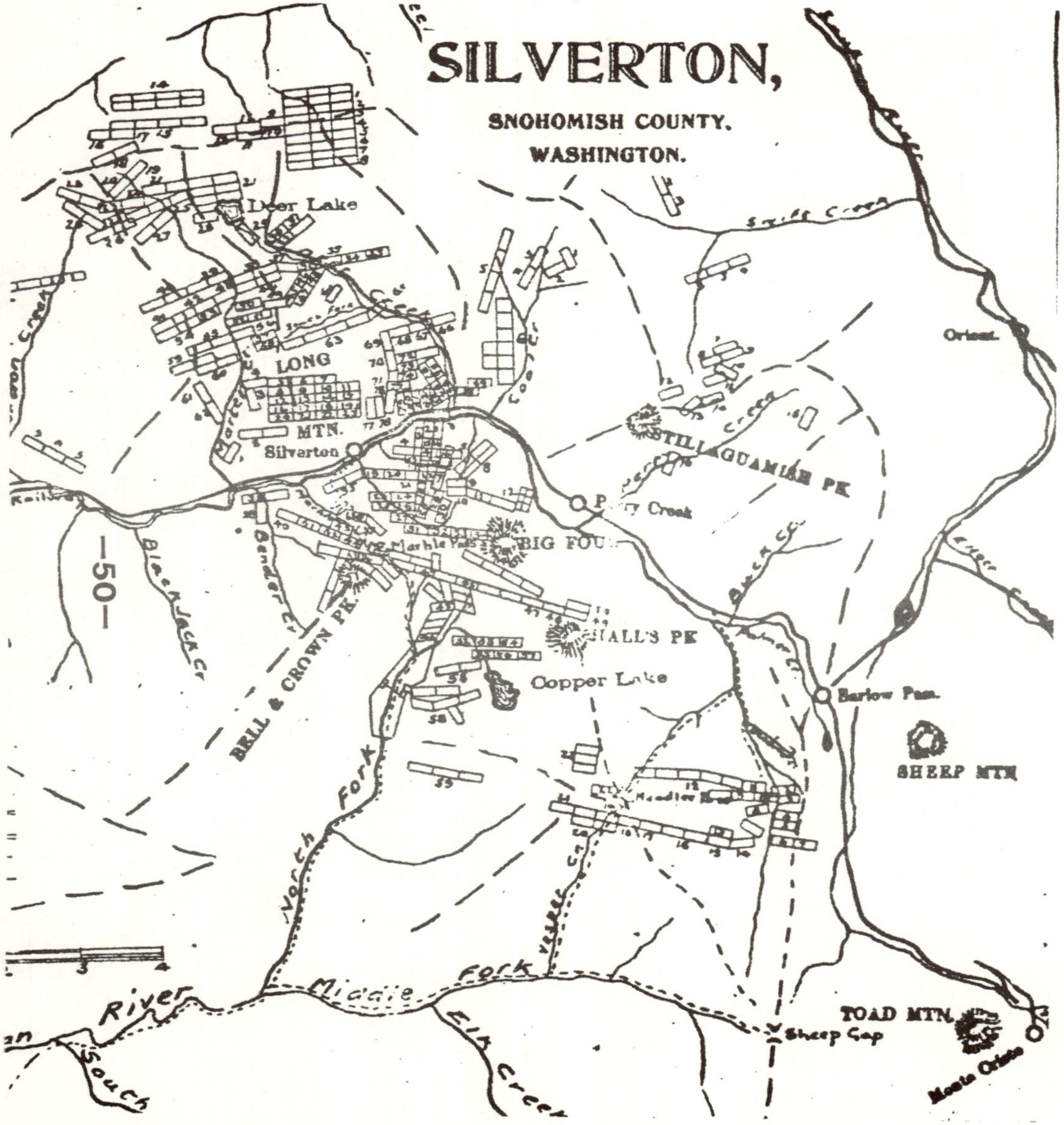
SILVERTON,
SNOHOMISH COUNTY.
WASHINGTON.
Deer Lake
LONG
MTN.
Silverton
STILLAGUAMISH PK
BIG FOUR
HALL'S PK
Copper Lake
BELL & CROWN PK
Barlow Pass
SHEEP MTN
Orient
TOAD MTN
Sheep Gap
Monte Cristo
North Fork
Middle Fork
River
South
Elk Creek
Black Jack Cr
Bender Cr
Marble Pass

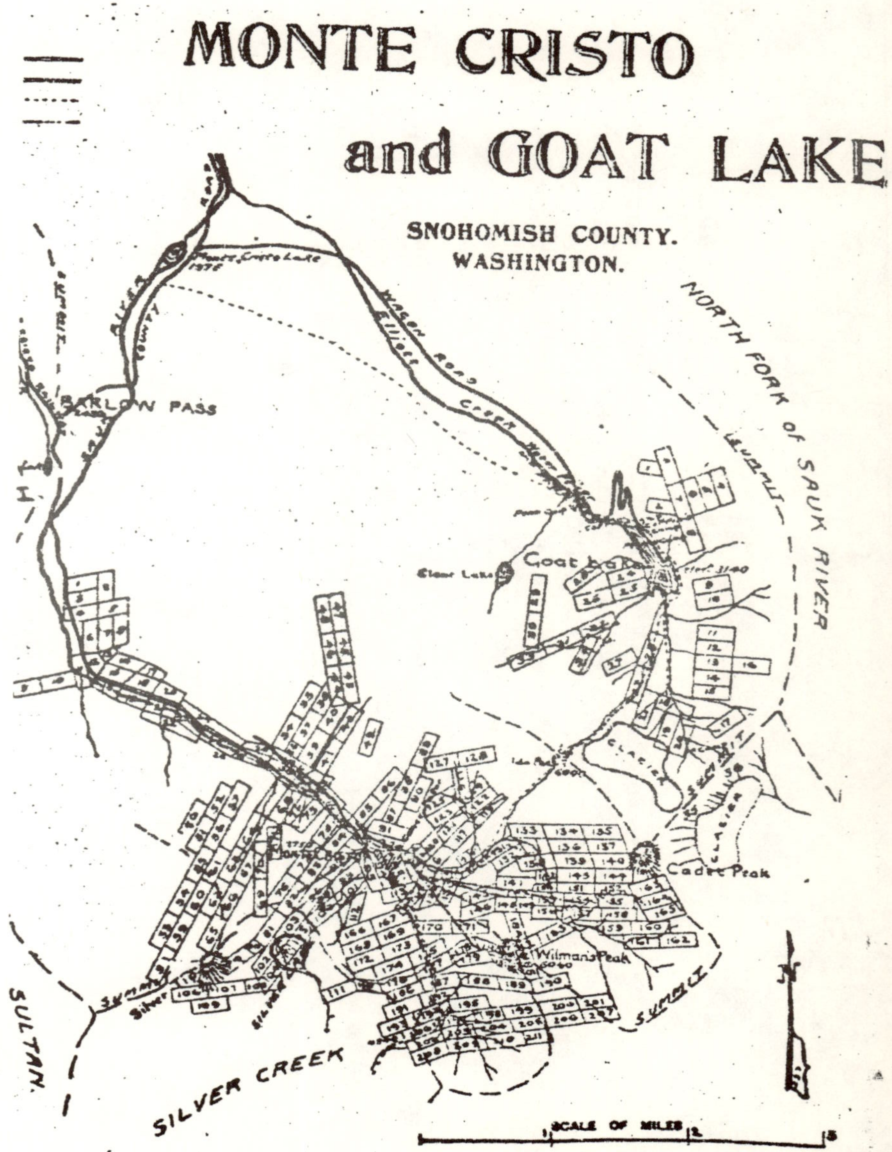
MONTE CRISTO
and GOAT LAKE
SNOHOMISH COUNTY.
WASHINGTON.
NORTH FORK OF SAUK RIVER
BARLOW PASS
Wagon Road
Elliott Creek
Goat Lake
Cadet Peak
Wilmans Peak
SILVER CREEK
SULTAN.
SCALE OF MILES

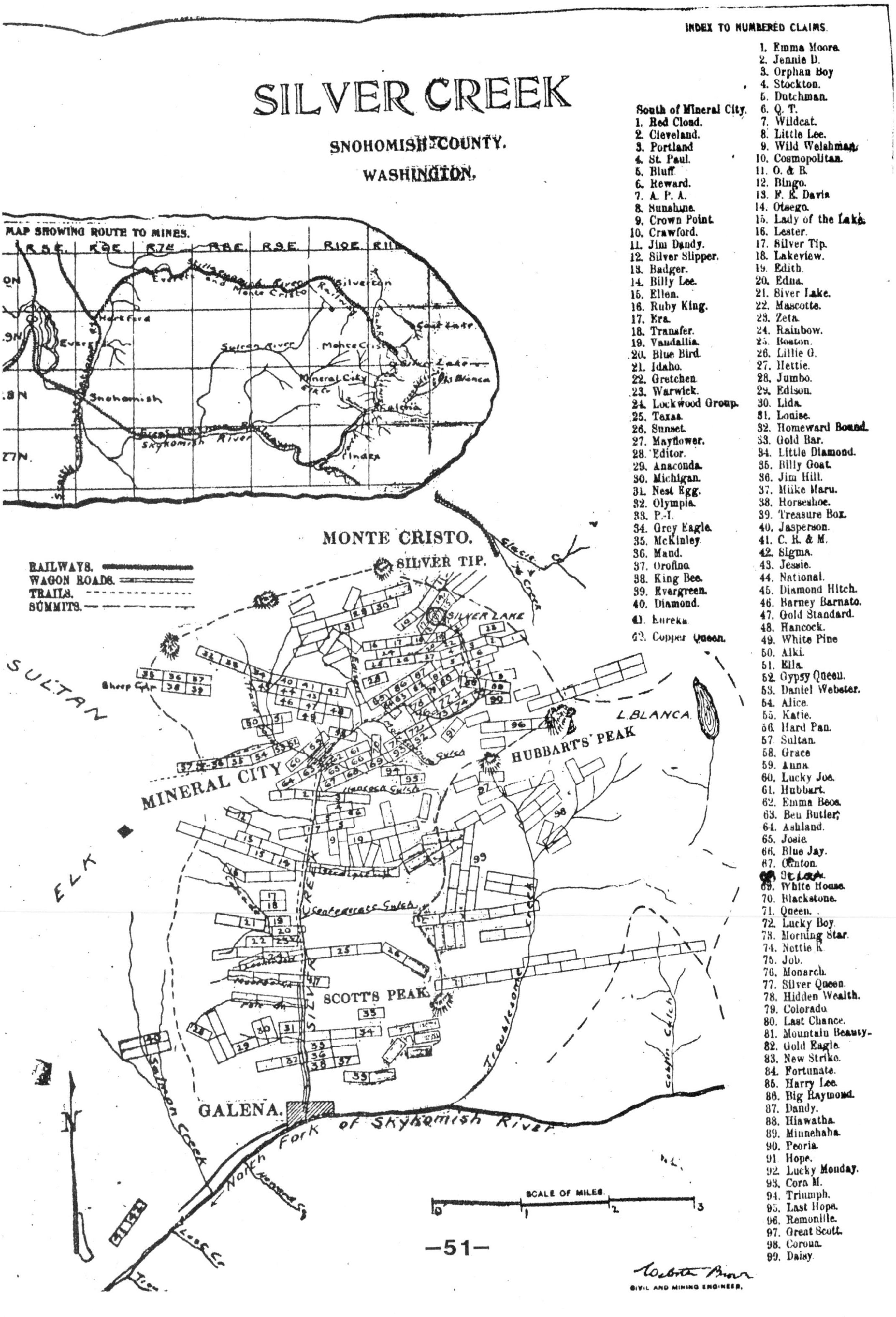
SILVER CREEK
SNOHOMISH COUNTY.
WASHINGTON.
MAP SHOWING ROUTE TO MINES.
Snohomish
Skykomish River
Sultan River
Monte Cristo
Mineral City
Silverton
Index
MONTE CRISTO.
SILVER TIP.
SILVER LAKE
RAILWAYS.
WAGON ROADS.
TRAILS.
SUMMITS.
SULTAN
ELK
MINERAL CITY
HUBBART'S PEAK
L.BLANCA
SCOTT'S PEAK
GALENA.
North Fork of Skykomish River
Salmon Creek
Troublesome Creek
Glacier Creek
Confederate Gulch
SCALE OF MILES.
0 1 2 3
South of Mineral City.
1. Red Cloud.
2. Cleveland.
3. Portland
4. St. Paul.
5. Bluff
6. Reward.
7. A. P. A.
8. Sunshine.
9. Crown Point
10. Crawford.
11. Jim Dandy.
12. Silver Slipper.
13. Badger.
14. Billy Lee.
15. Ellen.
16. Ruby King.
17. Era.
18. Transfer.
19. Vandalia.
20. Blue Bird.
21. Idaho.
22. Gretchen.
23. Warwick.
24. Lockwood Group.
25. Texas.
26. Sunset.
27. Mayflower.
28. Editor.
29. Anaconda.
30. Michigan.
31. Nest Egg.
32. Olympia.
33. P.-I.
34. Grey Eagle.
35. McKinley.
36. Maud.
37. Orofino.
38. King Bee.
39. Evergreen.
40. Diamond.
41. Eureka.
42. Copper Queen.
INDEX TO NUMBERED CLAIMS.
1. Emma Moore.
2. Jennie D.
3. Orphan Boy
4. Stockton.
5. Dutchman.
6. Q. T.
7. Wildcat.
8. Little Lee.
9. Wild Welshman.
10. Cosmopolitan.
11. O. & B.
12. Bingo.
13. F. K. Davis
14. Otsego.
15. Lady of the Lake.
16. Lester.
17. Silver Tip.
18. Lakeview.
19. Edith.
20. Edna.
21. Siver Lake.
22. Mascotte.
23. Zeta.
24. Rainbow.
25. Boston.
26. Lillie G.
27. Hettie.
28. Jumbo.
29. Edison.
30. Lida.
31. Louise.
32. Homeward Bound.
33. Gold Bar.
34. Little Diamond.
35. Billy Goat.
36. Jim Hill.
37. Mike Maru.
38. Horseshoe.
39. Treasure Box.
40. Jasperson.
41. C. R. & M.
42. Sigma.
43. Jessie.
44. National.
45. Diamond Hitch.
46. Barney Barnato.
47. Gold Standard.
48. Hancock.
49. White Pine
50. Alki.
51. Ella.
52. Gypsy Queen.
53. Daniel Webster.
54. Alice.
55. Katie.
56. Hard Pan.
57. Sultan.
58. Grace
59. Anna.
60. Lucky Joe.
61. Hubbart.
62. Emma Bees.
63. Ben Butler.
64. Ashland.
65. Josie.
66. Blue Jay.
67. Canton.
69. White House.
70. Blackstone.
71. Queen.
72. Lucky Boy.
73. Morning Star.
74. Nettie R.
75. Job.
76. Monarch.
77. Silver Queen.
78. Hidden Wealth.
79. Colorado.
80. Last Chance.
81. Mountain Beauty.
82. Gold Eagle.
83. New Strike.
84. Fortunate.
85. Harry Lee.
86. Big Raymond.
87. Dandy.
88. Hiawatha.
89. Minnehaha.
90. Peoria.
91. Hope.
92. Lucky Monday.
93. Cora M.
94. Triumph.
95. Last Hope.
96. Remonille.
97. Great Scott.
98. Corona.
99. Daisy.
CIVIL AND MINING ENGINEERS.

A mineralized zone in the Silverton area showed high assay values in gold, silver, copper, and other minerals. The railroad to Monte Cristo passed through here and helped in its development. The nature of these ores produced good placer areas along some parts of the Stillaguamish River. This is shown on Silverton Quadrangle.

The Sauk River at Darrington has produced placer gold. Gold Mountain, East of Darrington had lode mines producing gold, silver, copper and native mercury in cinnabar. Jumbo Mountain and White Horse Mountain had producing lodes. Buckeye Creek and Squire Creek should still produce some placer gold. Darrington $7\frac{1}{2}'$ and Silverton 15' Quadrangle show this area of Snohomish County.

Deer Creek, tributary of the North Fork Stillaguamish River, was placered near its mouth and on clay seams upstream. Difficulty in sluicing this clay made it unprofitable to work. This creek should still produce colors. Oso 15' Quadrangle shows the Deer Creek area of Skagit County.

An area of both placer and lode gold was discovered on the South Fork of the Stillaguamish River one to three miles above Granite Falls. This is shown on Granite Falls Quadrangle.

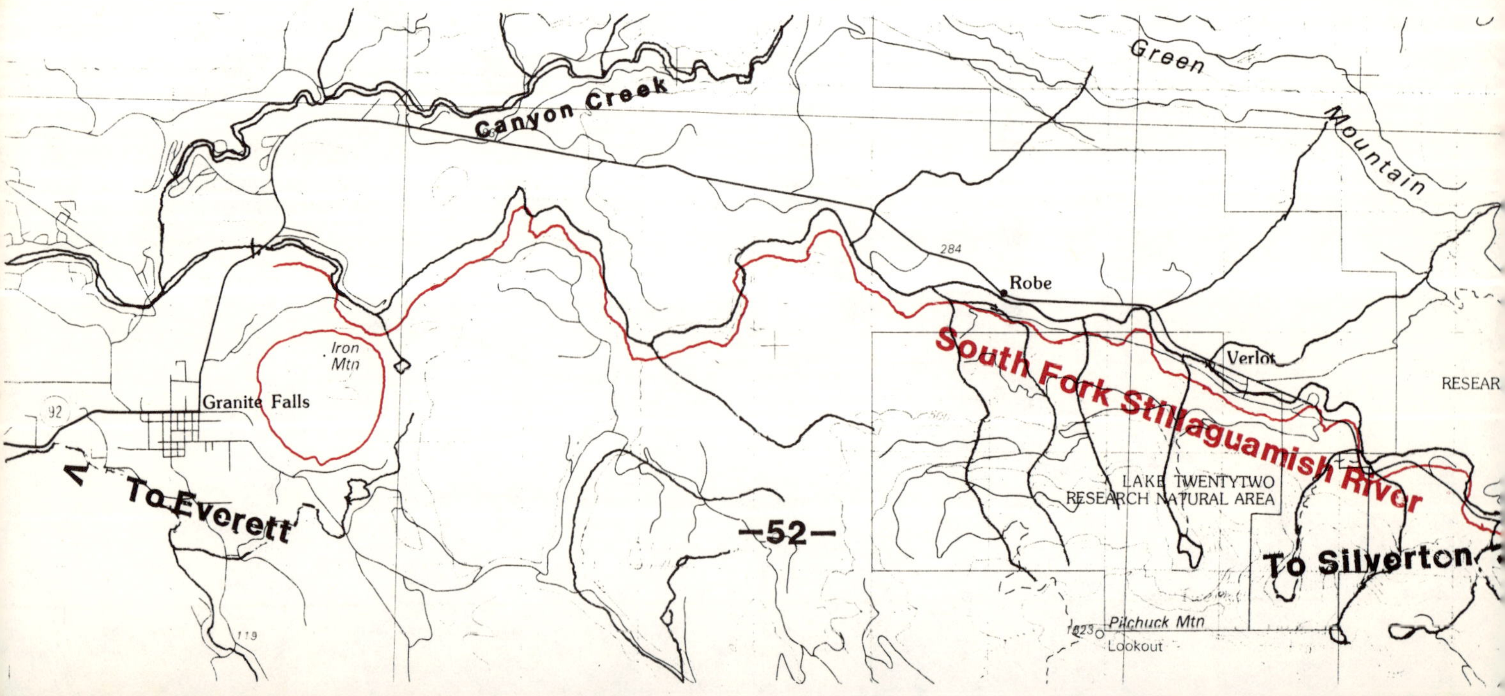

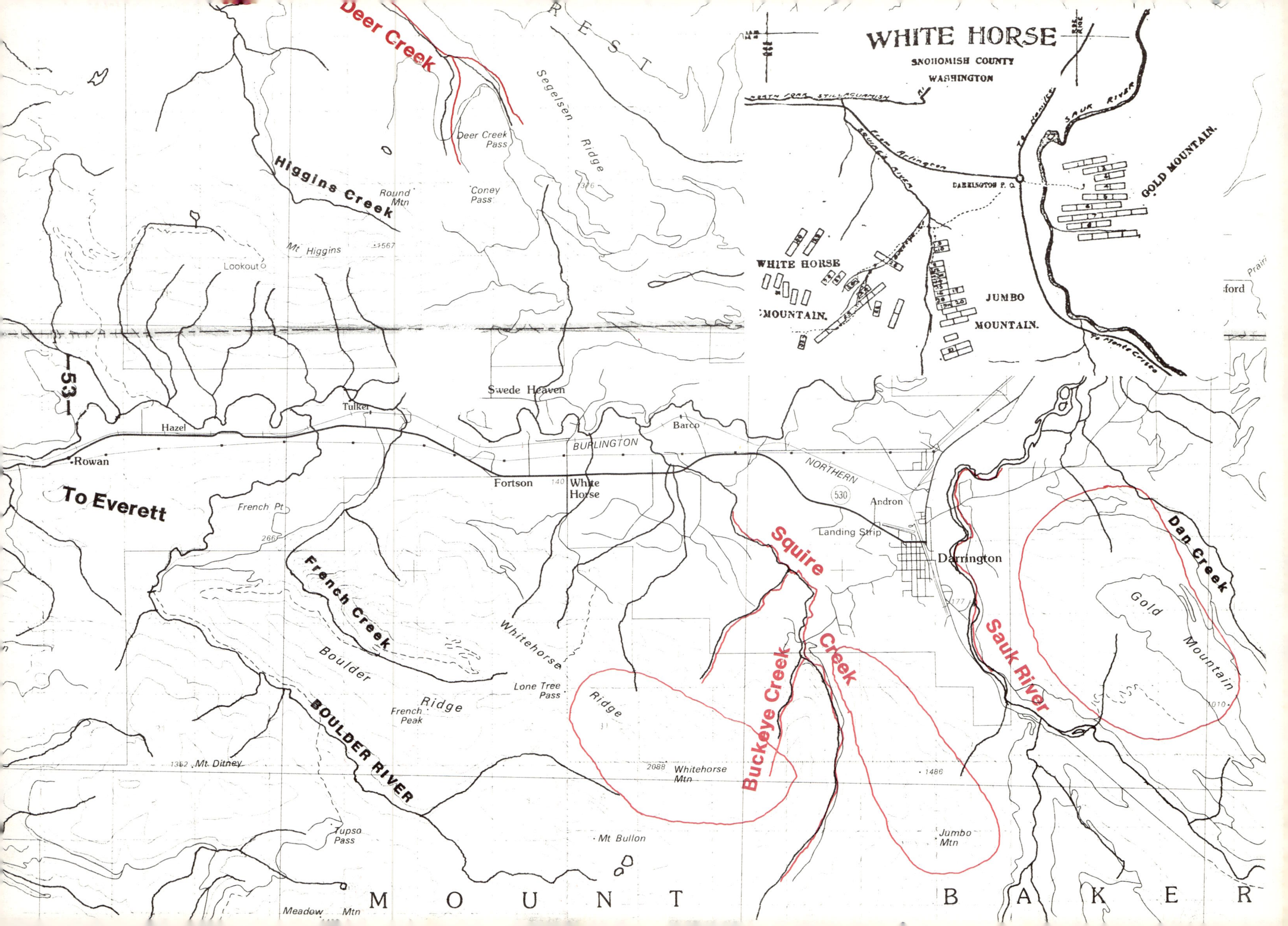
WHITE HORSE
SNOHOMISH COUNTY
WASHINGTON
GOLD MOUNTAIN.
JUMBO MOUNTAIN.
WHITE HORSE MOUNTAIN.
DARRINGTON P. O.
SAUK RIVER
Deer Creek
Deer Creek Pass
Segelsen Ridge
Higgins Creek
Round Mtn
Coney Pass
Mt Higgins
Lookout
Swede Heaven
Tulker
Hazel
Rowan
To Everett
Barco
BURLINGTON
NORTHERN
Fortson
White Horse
530
Andron
Landing Strip
Darrington
French Pt
French Creek
Boulder Ridge
Whitehorse
Lone Tree Pass
French Peak
BOULDER RIVER
Mt. Ditney
Squire Creek
Buckeye Creek
Whitehorse Mtn
Ridge
Jumbo Mtn
Mt Bullon
Tupso Pass
Meadow Mtn
Sauk River
Dan Creek
Gold Mountain
MOUNT BAKER

A mineralized area near the headwaters of the North Fork Snoqualmie River has produced considerable gold. Lodes in this area produce gold, silver, copper, lead, and other minerals. Streams draining all directions from this region have produced placer gold.

Coney Creek and West Fork Miller River draining into the Skykomish River have produced placer gold. Kimball Creek and Money Creek also drain this mineral zone. Gold has also been reported on the Tolt River. Mount Si 15' and Grotto 7½' Quadrangles show this part of King County.

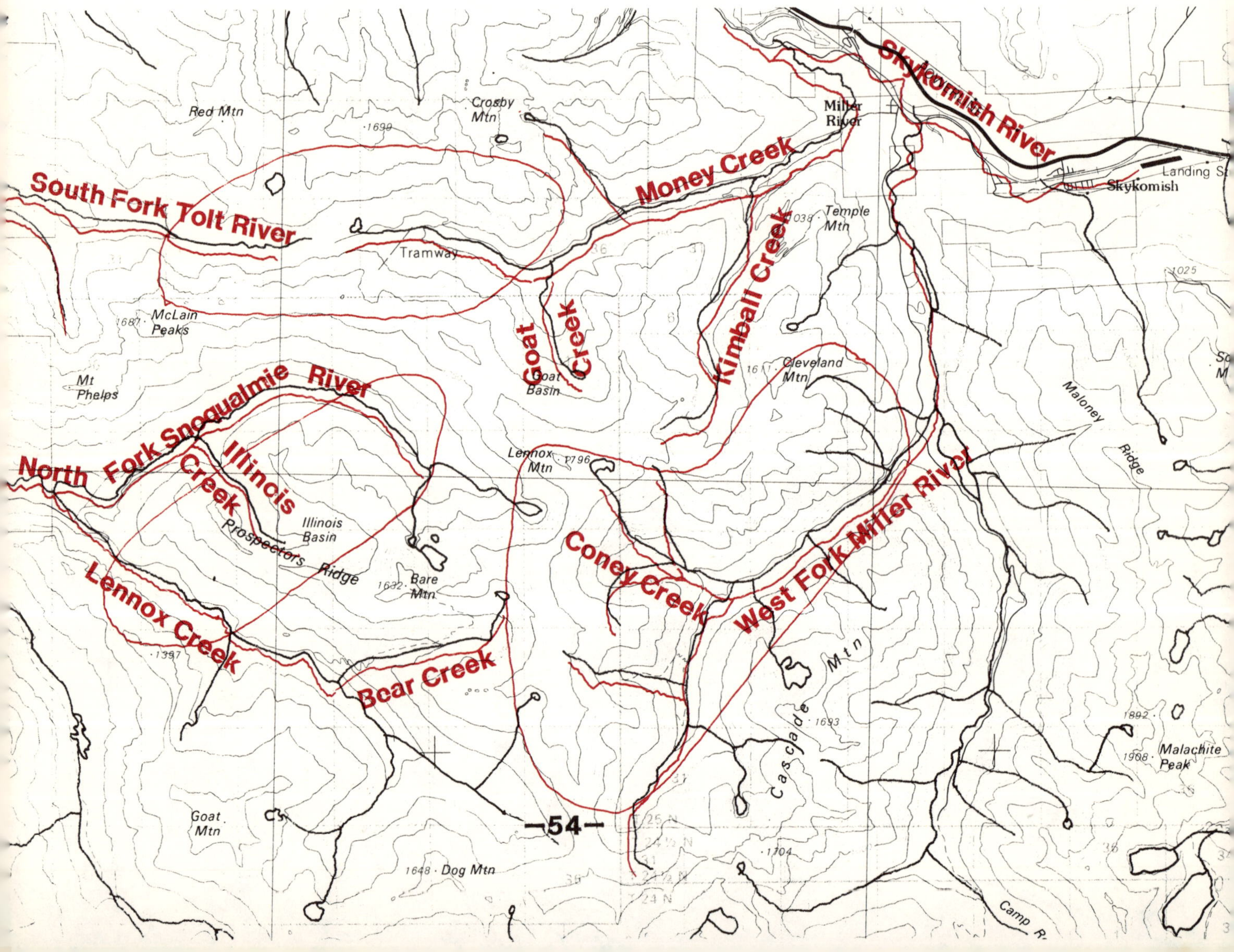

Miller River, Money Creek and Buena Vista,

KING COUNTY, WASHINGTON.

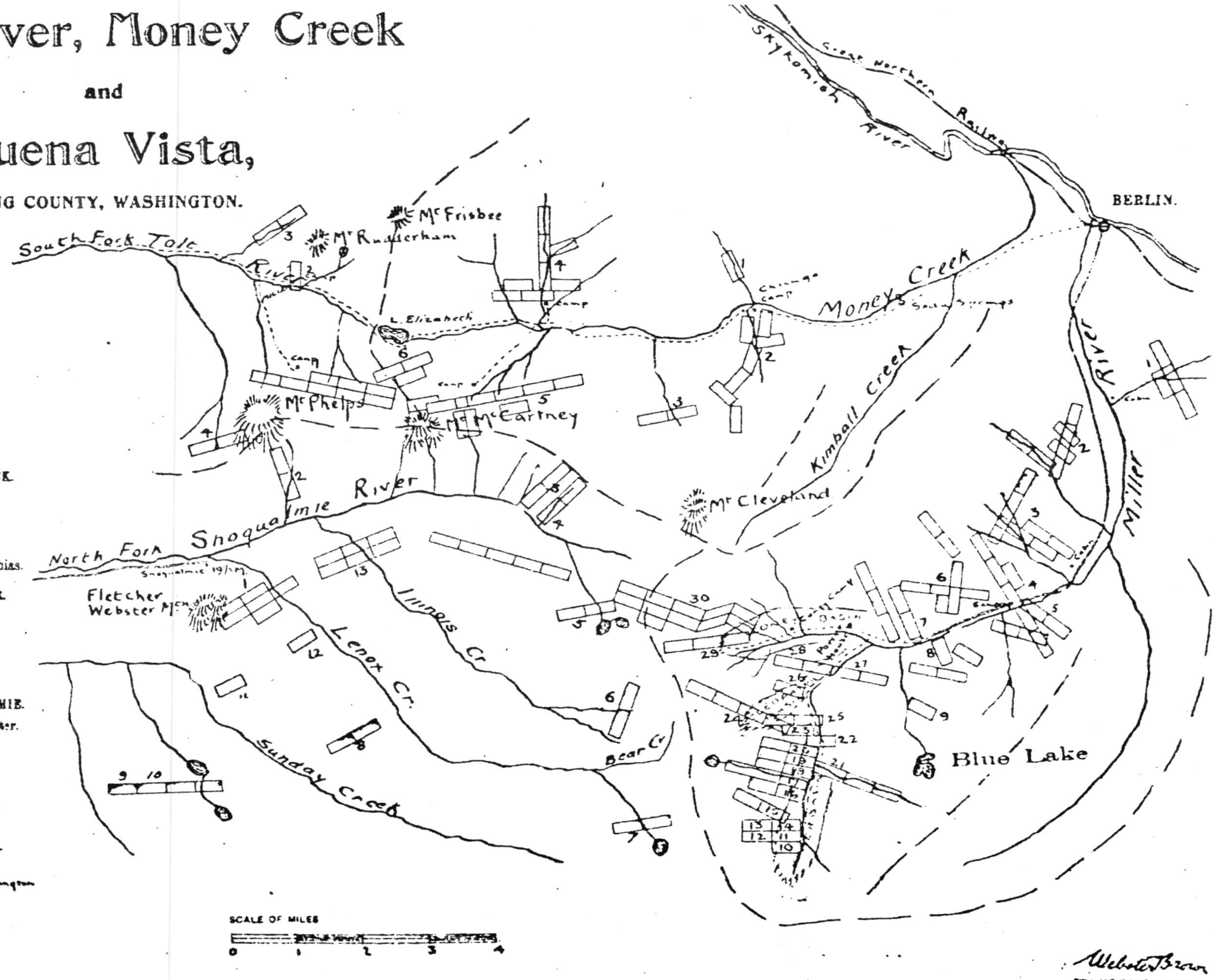

Railways.
Wagon Roads.
Trails.
Summit Lines.

INDEX TO NUMBERED CLAIMS.

MILLER RIVER.

1. Mono.
2. McKinley.
3. S. S. S.
4. Lynn.
5. Belle.
6. Little Una.
7. War Eagle.
8. Jay Hawker.
9. Mountain Lion.
10. Blucher.
11. Highlander.
12. Mountain Goat.
13. Captain.
14. Easter.
15. Clara K.
16. Great Northern.
17. Bobtail.
18. Grand Central.
19. Le Roy.
20. Washington.
21. Seattle.
22. Aces Up.
23. Lucky Jim.
24. Cleopatra Group.
25. Minnecadusa.
26. Baltimore.
27. U. P.
28. Condor.
29. Coney Group.
30. Brooklyn.

MONEY CREEK.

1. Red Coat.
2. Chicago.
3. Pinto.
4. San Francisco.
5. Apex.
6. Damon and Pythias.

TOLT RIVER.

1. Black Chief.
2. Mammoth.
3. Baltimore.
4. Lady Belle.

NORTH FORK SNOQUALMIE.

1. Fletcher Webster.
2. Red Cloud.
3. North Fork.
4. Betsy Ross.
5. Paradise.
6. Monitor.
7. Copper Queen.
8. Bunker Hill.
9. Illinois.
10. West Virginia.
11. Bay View.
12. Alliance.
13. Belle of Washington.

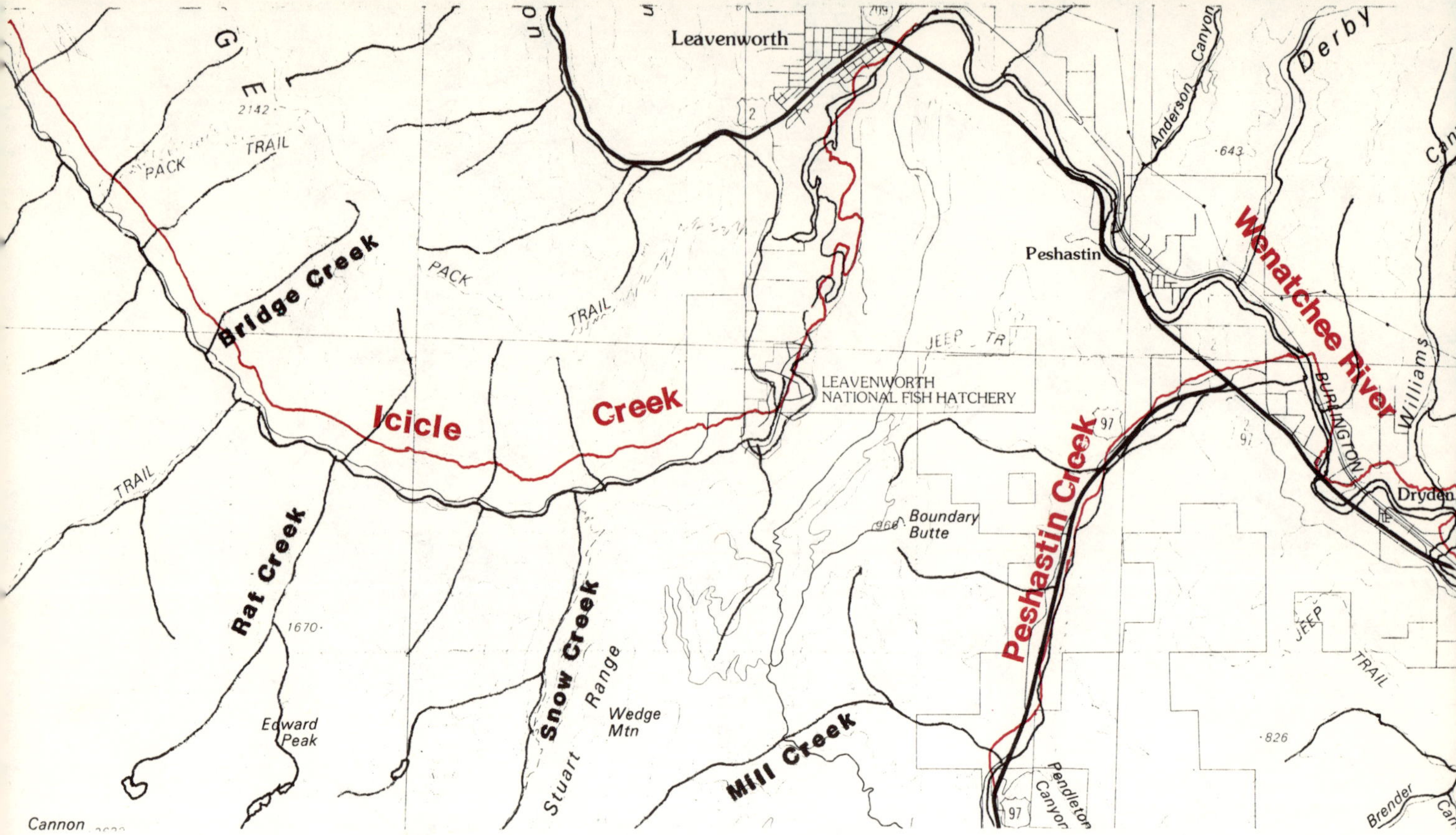

Gold has been found on the Wenatchee River from below Tumwater Canyon to Leavenworth. Icicle Creek has produced considerable gold, but good locations are scattered. From the mouth of Peshastin Creek to Dryden, the Wenatchee River will produce a few colors. These areas are shown on Leavenworth Quadrangle, Chelan County.

Deep Creek, near Lake Wenatchee, has produced placer gold. Plain $7\frac{1}{2}$' Quadrangle shows this creek.

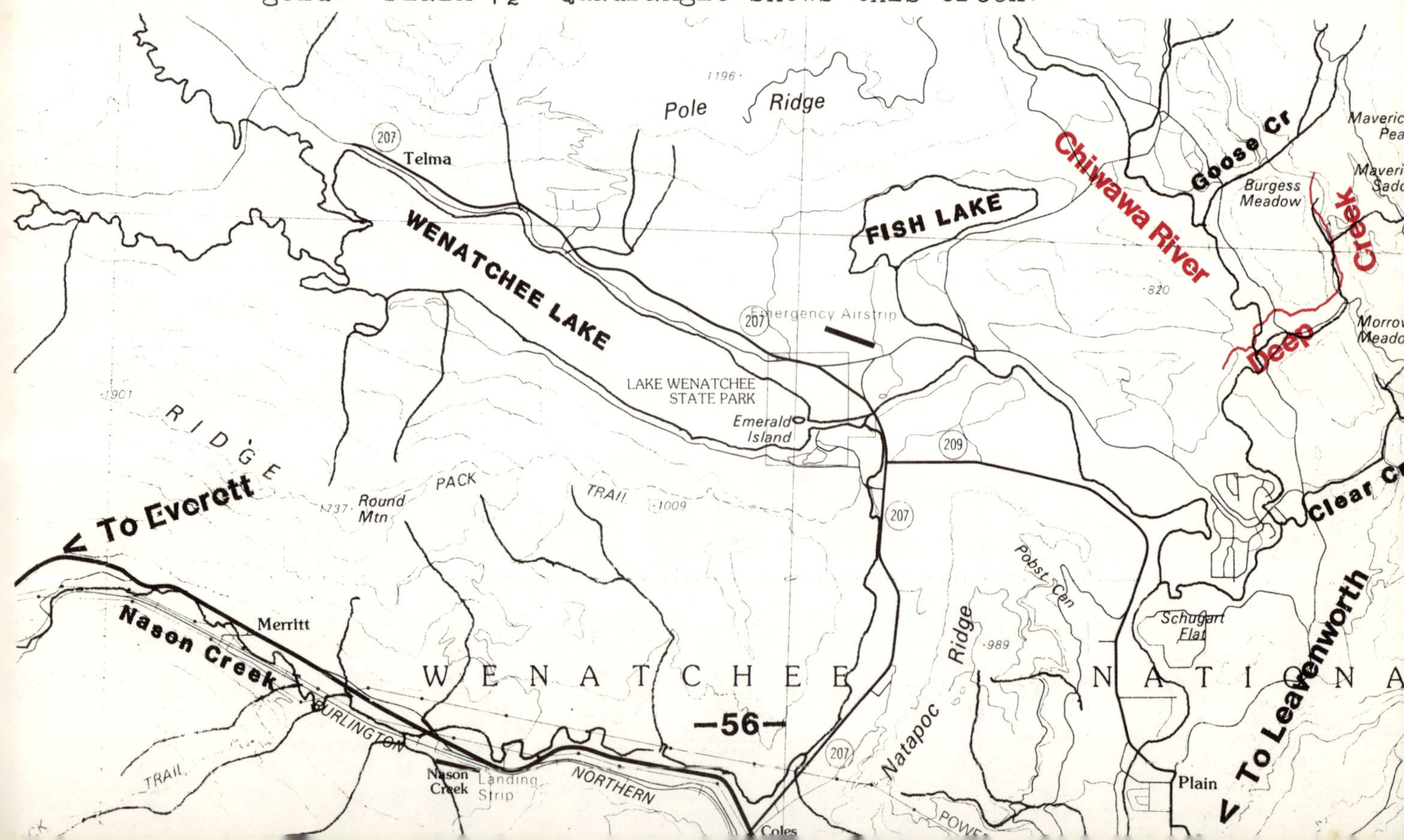

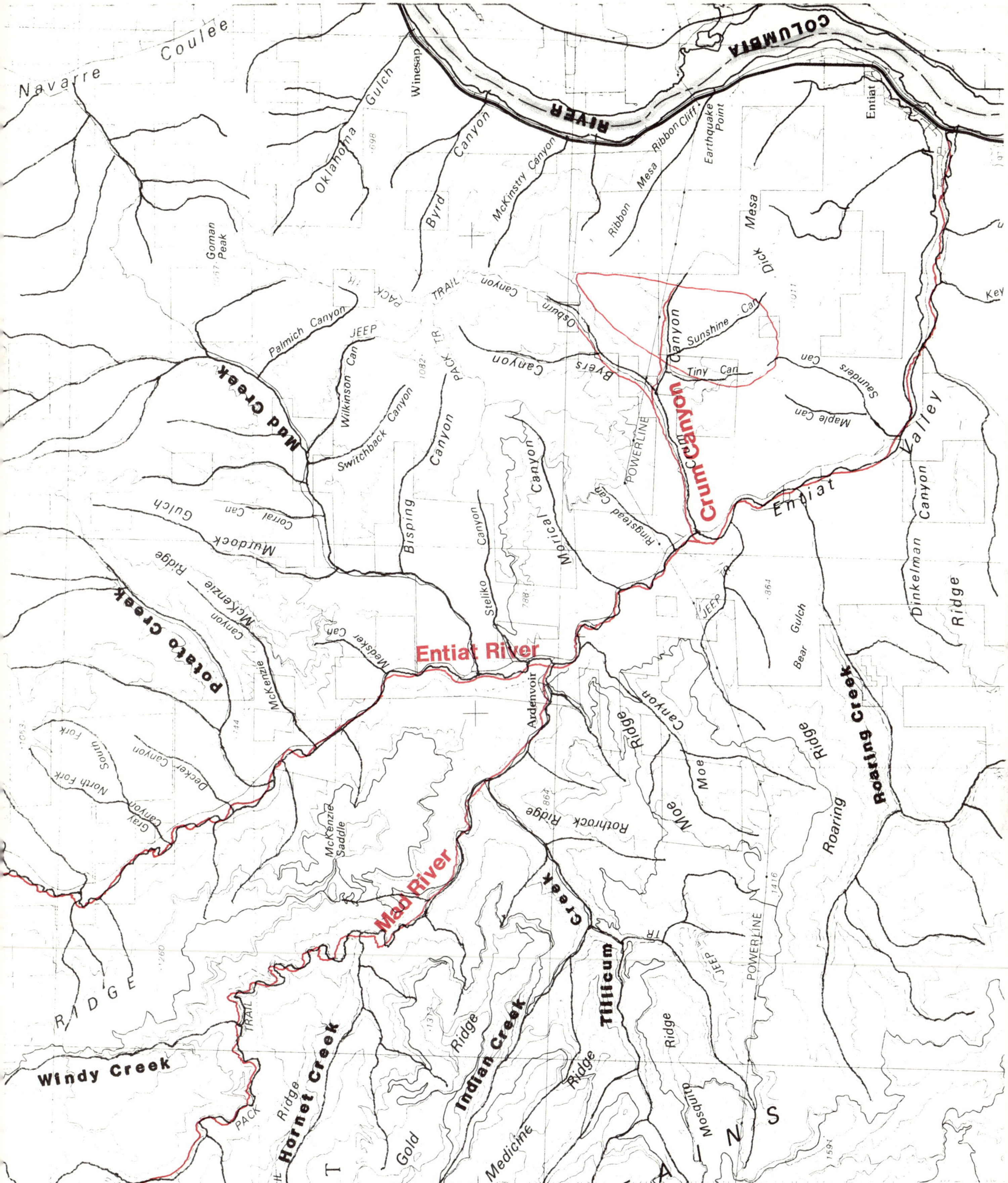

Lode mines in Crum Canyon have produced gold. Crum Canyon is a tributary of the Entiat River. The Entiat River and the Mad River have many placers. These areas are shown on Entiat, Ardenvoir, Chumstick Mountain, Baldy Mountain, Tyee Mountain, Sugarloaf Peak, Brief, Silver Falls 7½' and Lucerne 15' Quadrangles.

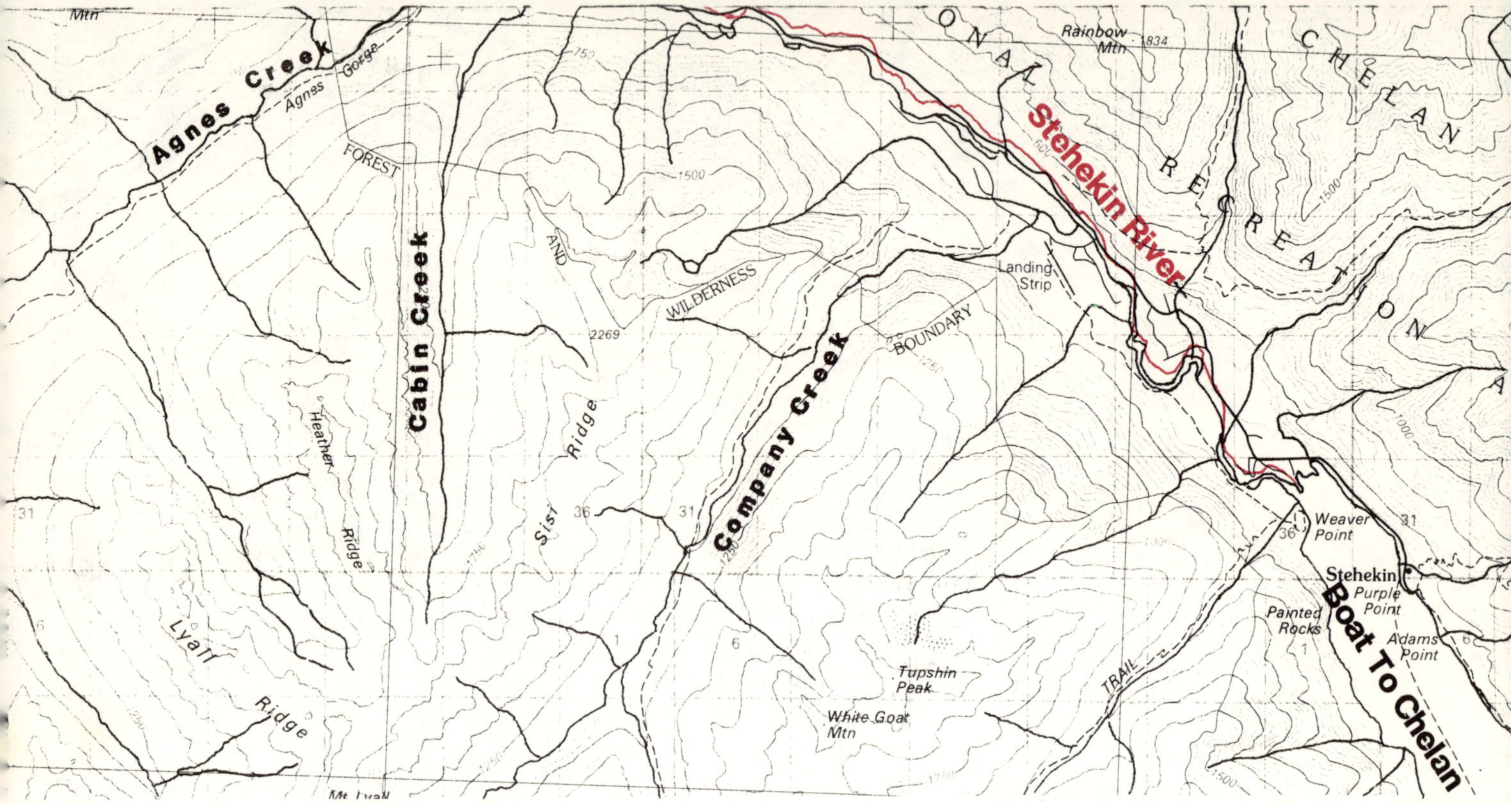

Some placer gold has been found on beaches in several places on Lake Chelan. Many creeks along the upper part of the lake will produce some gold. Railroad Creek has several good placer locations. The Stehekin River has had placers for several miles above its mouth. The upper Lake Chelan area is shown on Lucerne 15' and Mt. Lyall 7½' Quadrangle.

Although discovered in 1896, the Holden Mine on Railroad Creek was not developed until 1938. During its twenty year life it produced 10.6 million tons of gold, copper, and silver ore valued at $66,500,000.00. This mine employed 450 men who dug 250,000 feet of tunnel on a mineralized zone up to 75 feet wide. This fabulous mine was a valuable asset to the depression days economy. All this production was shipped by barge to Chelan. The upper Lake Chelan area is only accessible by boat. This property is now operated as a church camp. Holden Quadrangle shows the Holden Mine and part of Railroad Creek.

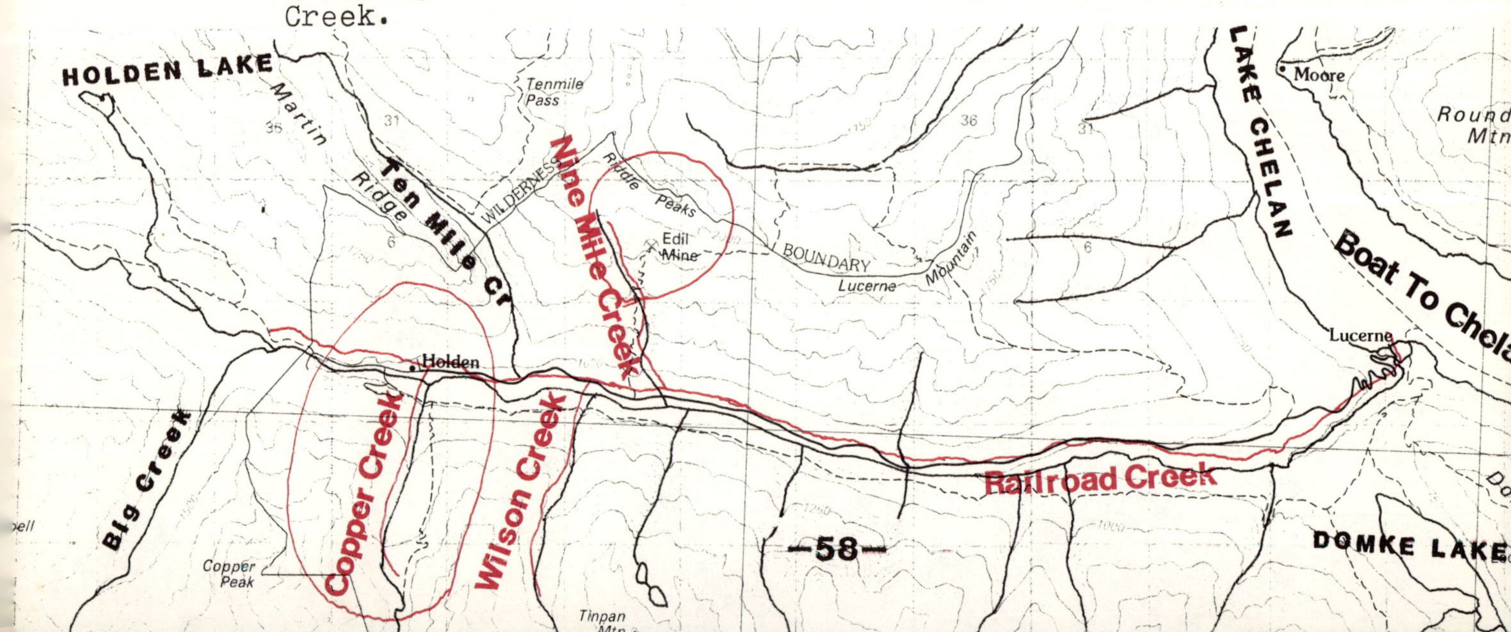

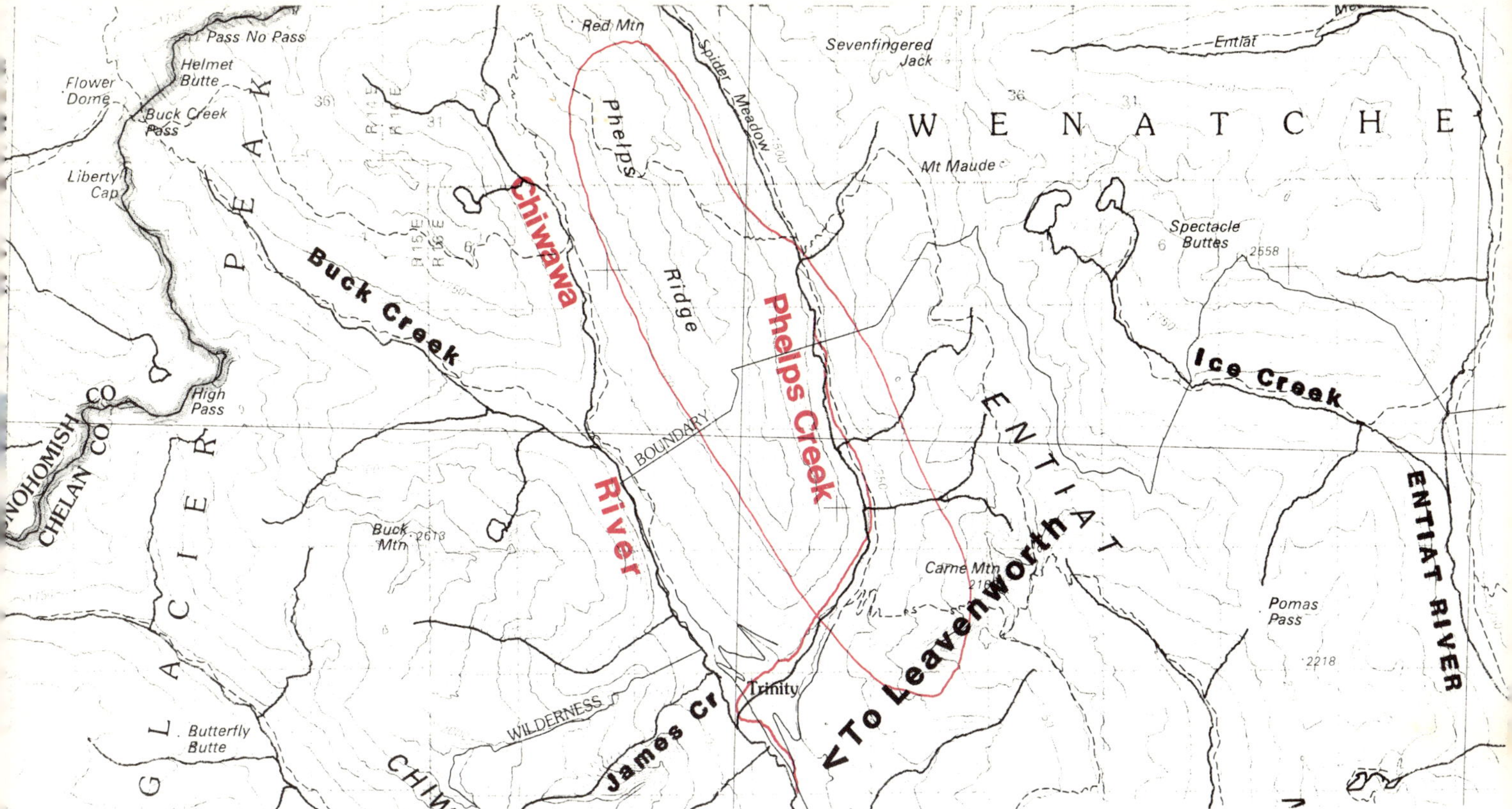

The upper Chiwawa River has several lode mines that show high assay values. Phelps Creek should produce placer gold. This area of Chelan County is also shown on Holden Quadrangle.

The Chelan Butte area has produced lode gold. The Chelan River and the Columbia River below its mouth was placer mined by Chinese miners. Streams draining Chelan Butte may have been the source of some of this placer gold. Chelan and Chelan Falls Quadrangles show this area.

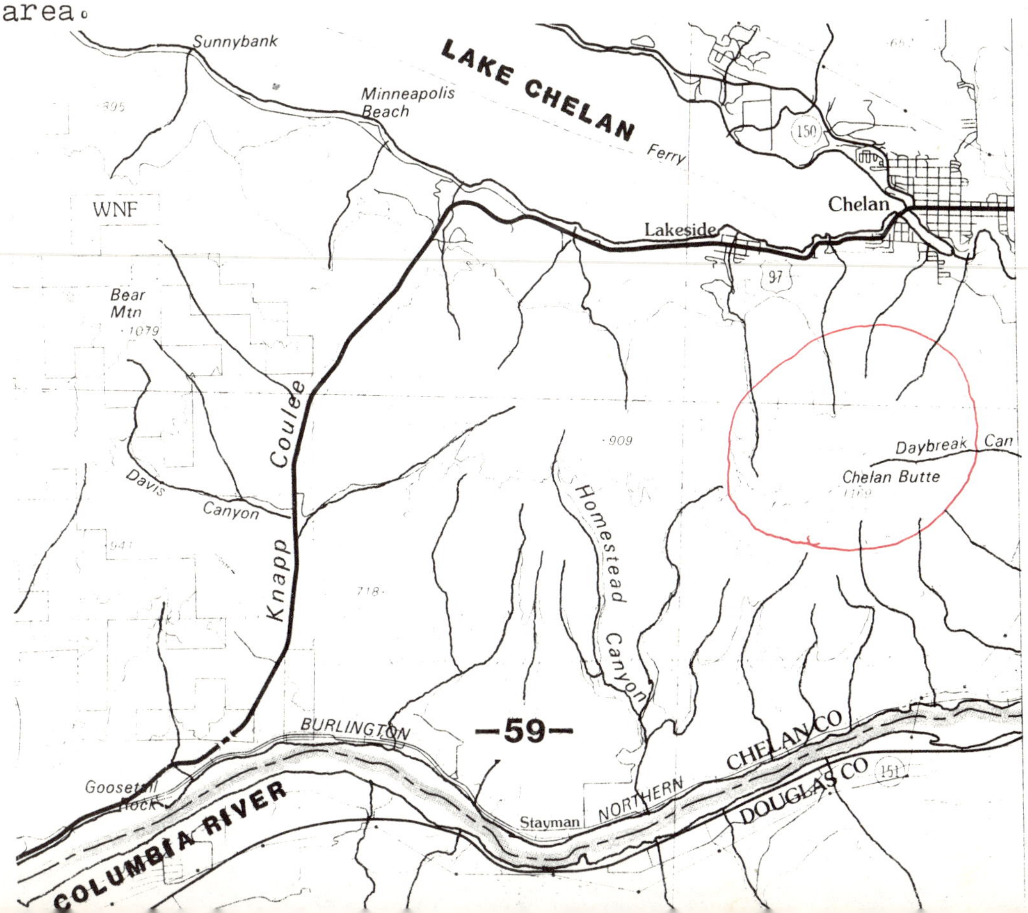

The Cascade River drainage, tributary of the Skagit River, has a history of mining dating from 1887 when a troop of soldiers crossing Cascade Pass discovered rich mineral outcroppings. This mineralized area extends for several miles and includes the headwaters of the Stehekin River and Thunder Creek. Gold was one of the lesser values in this ore, but showed some good assays. Some of this area is now within the North Cascades National Park. See Cascade Pass Quadrangle.

Some placer gold has been found on the Cascade River between Lookout Creek and Sibley Creek. Day Creek has also produced. See Marblemount Quadrangle for this area of Skagit County.

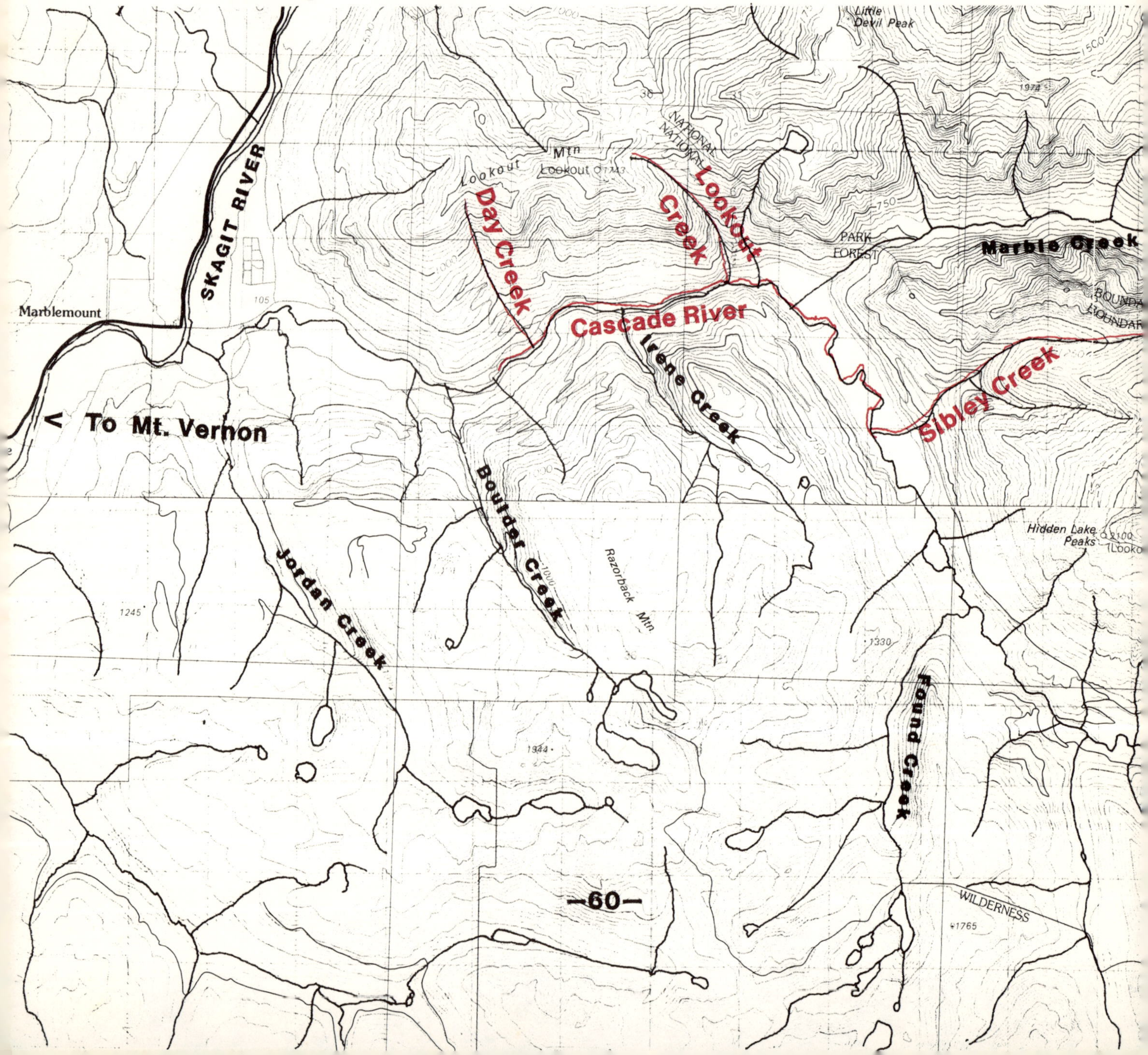

A mineralized zone starts near Nooksack Falls on the North Fork Nooksack River and includes all the upper drainage of the river and its tributaries. Several lodes in this area have been good producers of gold and other minerals. Many streams have had good placers.

Wells Creek has had placers for several miles of its length. Bar Creek may have some good spots. Low grade ore was found on Many Sisters Creek. This creek may produce colors. These are shown on Mt. Baker Quadrangle.

Placers have been worked on the Nooksack River from the town of Shuksan upstream to its headwaters. Ruth Creek has placers to its headwaters and lodes have produced near Hannegan Pass which assayed several ounces of gold per ton of ore.

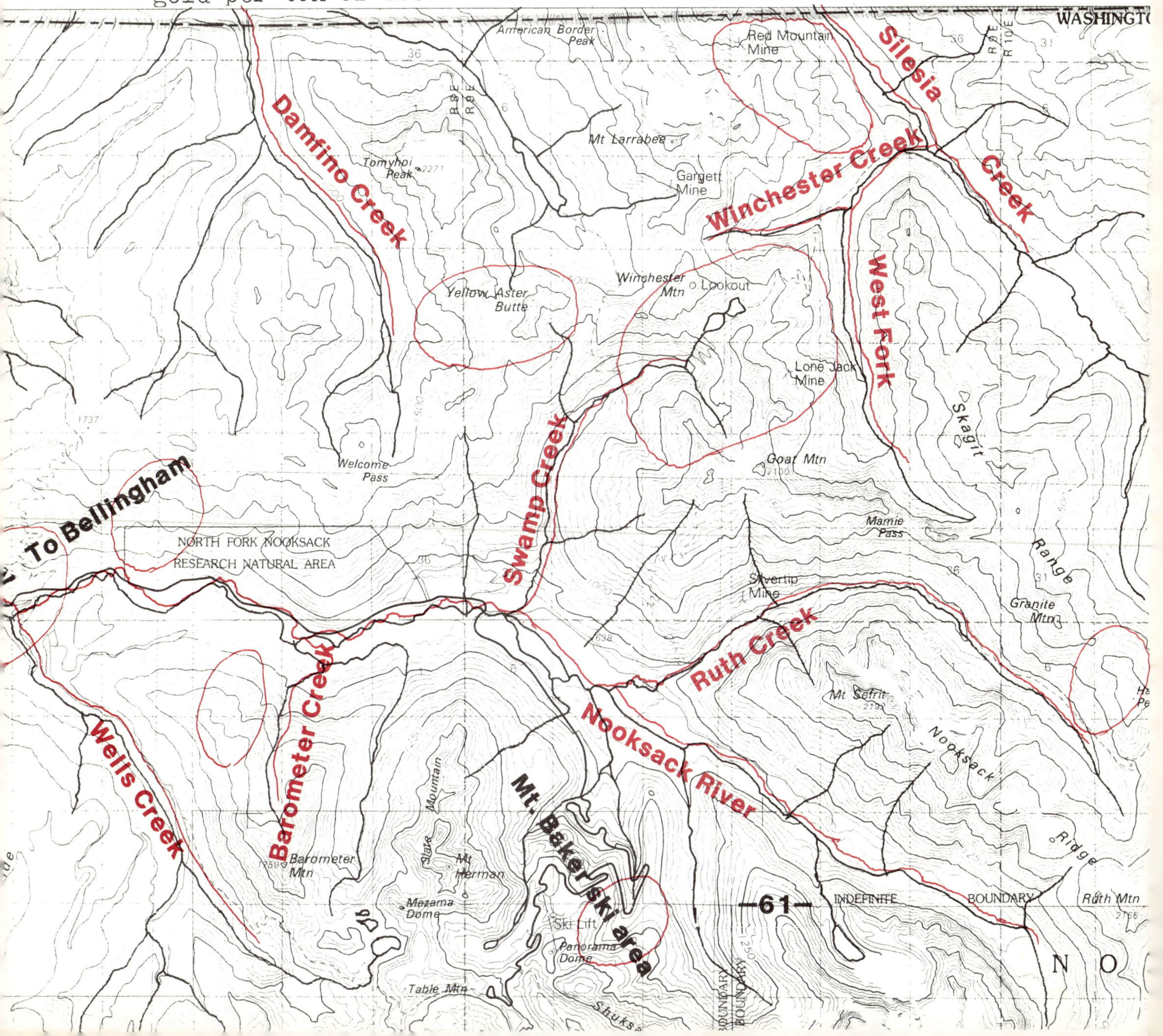

Swamp Creek, flowing into the Nooksack River from the north near Shuksan, has placers to its headwaters. Several rich lode mines have been worked on the ridge between Goat Mountain and Winchester Mountain.

An 1897 map of this area shows five cities in this district. Gold Hill was on the Nooksack River near the present location of Shuksan. Trail City was just upstream at the mouth of Swamp Creek. Union City was on Swamp Creek at the mouth of Willow Creek. Wilson's Townsite was on the Nooksack River at the mouth of Ruth Creek and Gold City was two and one half miles up Ruth Creek.

Many of the mines in this area are at high elevation and are in an area of heavy winter snows. Short working seasons and bad weather caused many labor problems. One mine is reported to have had three crews, one coming, one going, and one working.

Streams flowing northeast from the ridge in the vicinity of Skagway Pass carry placer gold. Silesia Creek, West Fork Silesia Creek, and Winchester Creek have all produced gold. Lode mines above these streams have produced considerable gold.

One of the richest lodes in this area was located just south of the Canadian border. The BoundaryRed Mountain and the Gold Basin Mine produced several hundred thousand dollars of gold. These mines may not have been fully developed because of the severe weather and short working seasons in this remote area. Production from the Gold Basin Mine was from float quartz only. The source of this rich float was never located.

Damfino Creek has produced placer gold. Many of the tributaries of this stream should also show colors.

Swift Creek and its tributaries flowing into Baker Lake and Baker River above Baker Lake have all produced glacer gold.

See Mt. Shuksan Quadrangle for this section of Whatcom County.

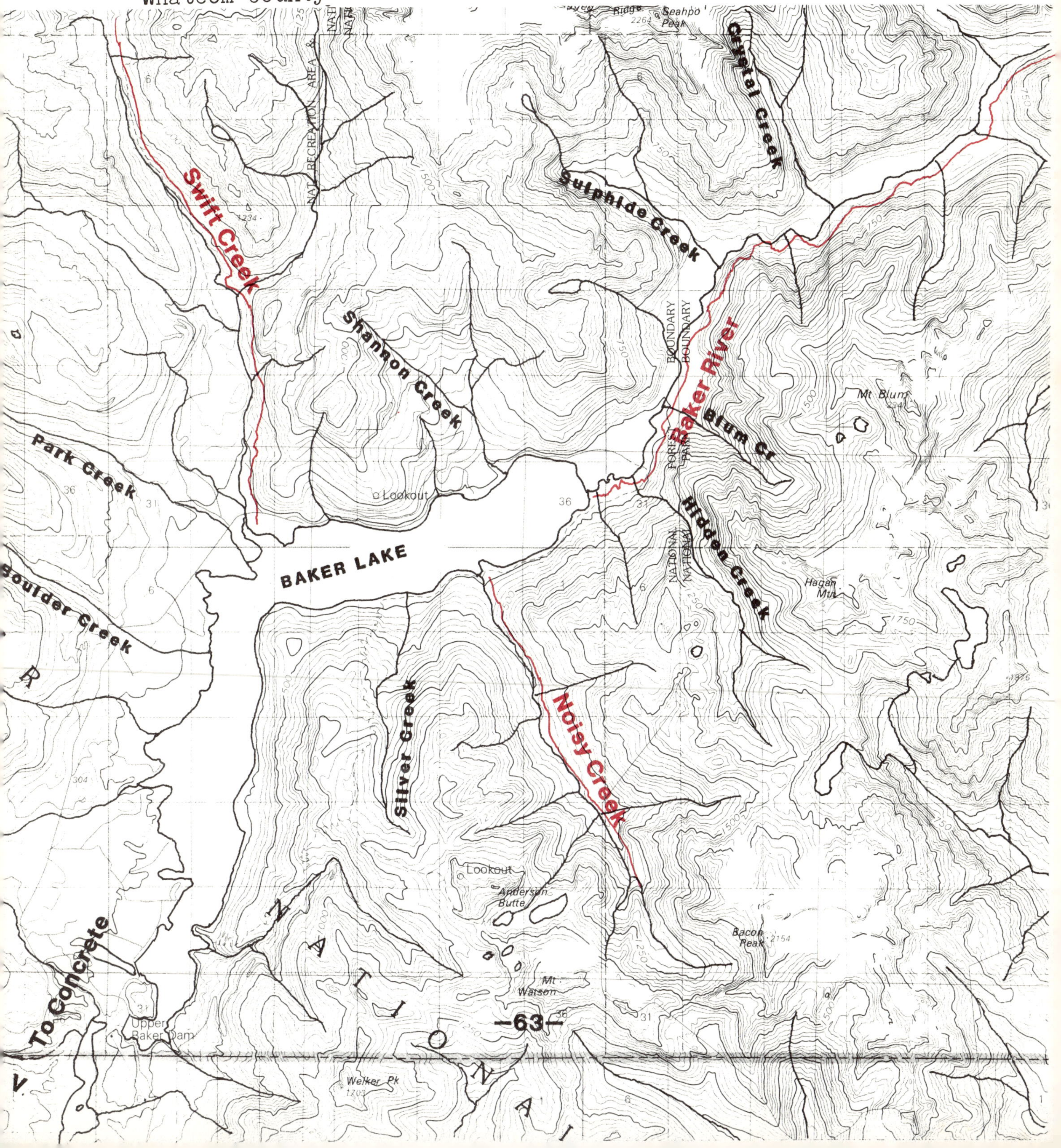

SLATE CREEK MINING DISTRICT

The first discovery of placer gold west of the Cascade Mountains was made on Ruby Creek or one of its tributaries. Ruby Creek is a tributary of the Skagit River and flows into Ross Lake just above Ross Dam. Some placers on Ruby Creek and the Skagit River are now flooded.

The Ruby Creek drainage covers a mineral rich area on the west slope of the Cascade Mountains in the vicinity of Harts Pass. Almost every stream in the area has produced some placer gold. Coarse gold nuggets were common in the early mining days. Over $100,000.00 is reported to have been recovered from placers soon after the first discovery. Two thousand five hundred men stampeded in on the rush the first season.

The present North Cascades Highway runs along Ruby Creek and Granite Creek to Rainy Pass. Granite Creek has been placered for several miles above Ruby Creek. Canyon Creek, Boulder Creek, and Mill Creek all have had placers. Slate Creek has several miles of good placers and also gives its name to this mining district.

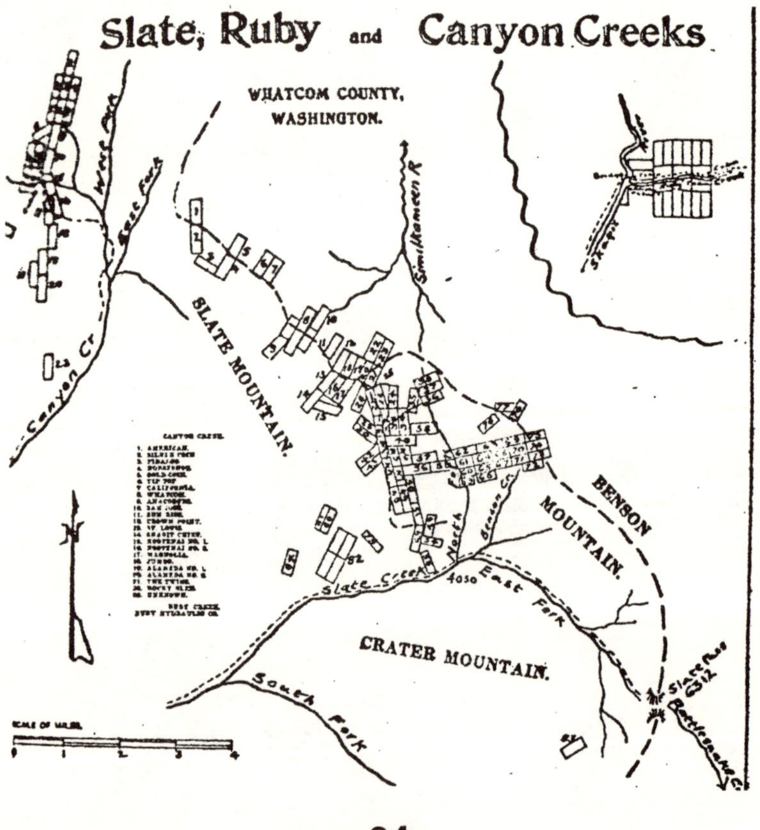

The New Light Mine was one of the first lode claims staked in the Slate Creek Mining District. It has had almost continuous production since 1895 and is still producing gold. This is only one of many lode mines in this area.

This area is shown on Ross Dam, Crater Mountain, Azurite Peak, Slate Peak, Shull Mountain and Pasayten Peak 7½' Quadrangles.

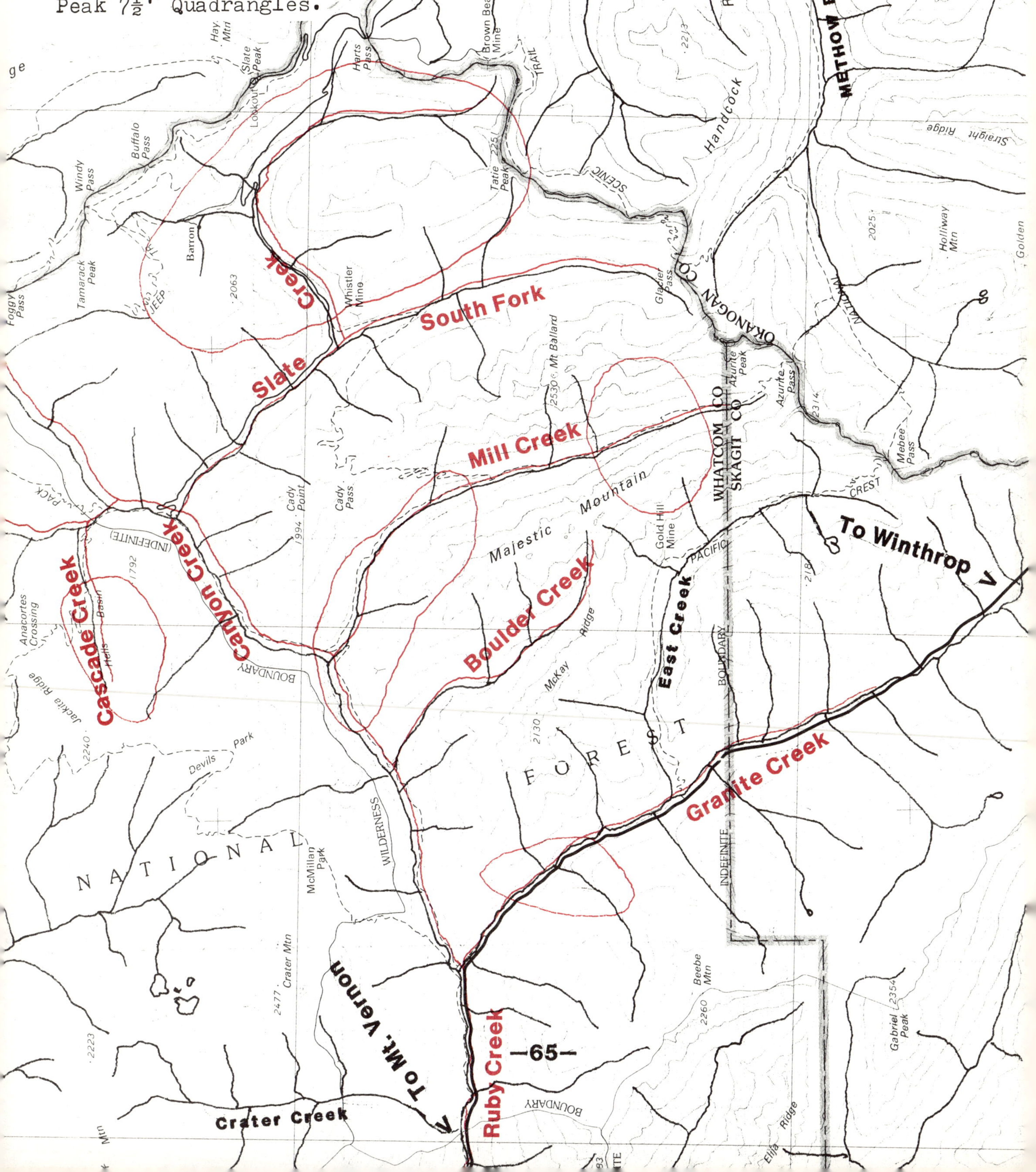

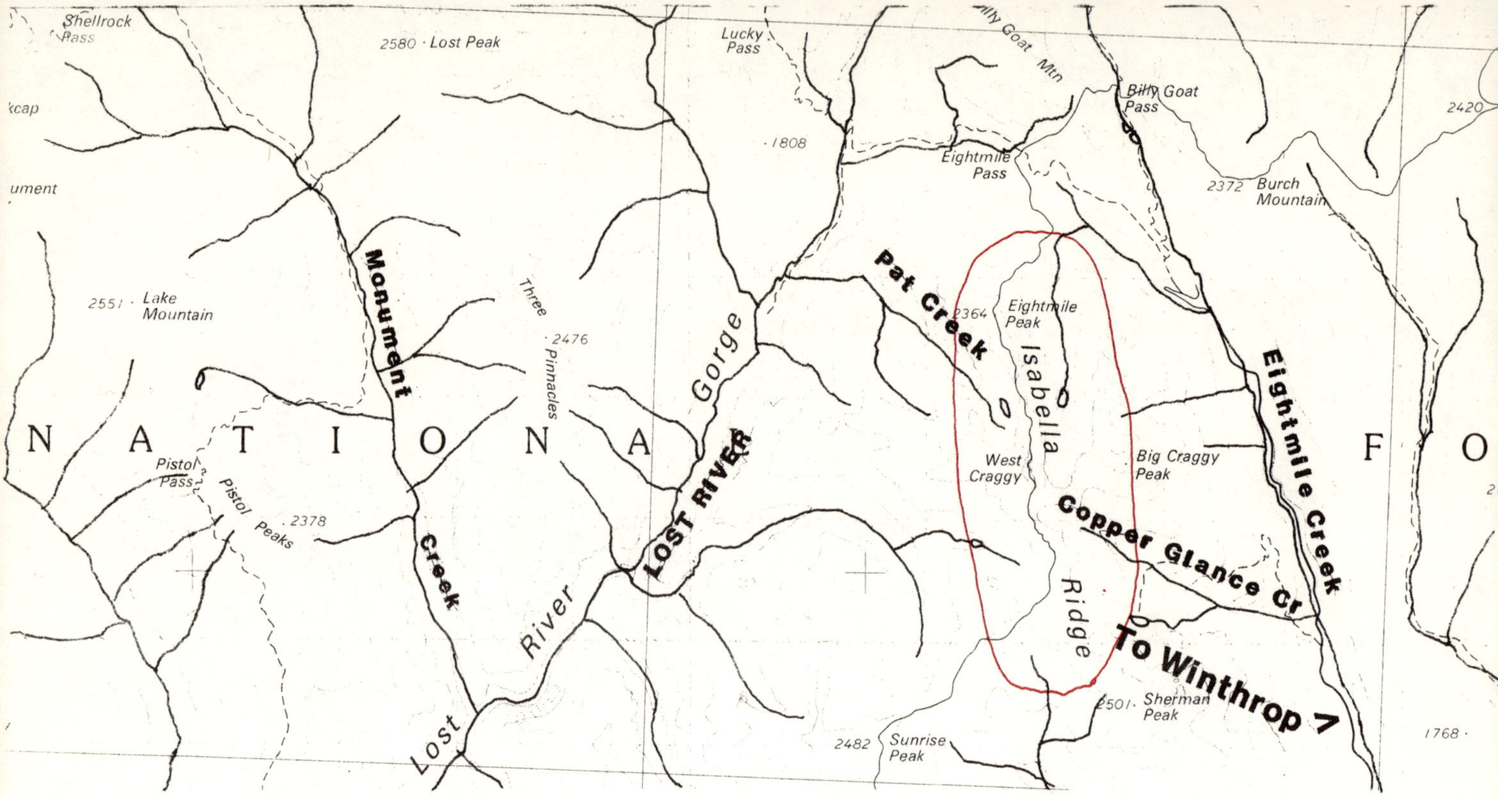

Lode gold was found on Isabella Ridge. Several claims have been worked in this area. Some had assays as high as one and one half ounces per ton of ore. See Billy Goat Mountain Quadrangle for this area of Okanogan County.

Lode mines have been workediin the vicinity of Mazama. Several claims have been staked along the Goat Wall, on the cliff north of Mazama and up Goat Creek near the mouth of Whiteface Creek. See Mazama Quadrangle.

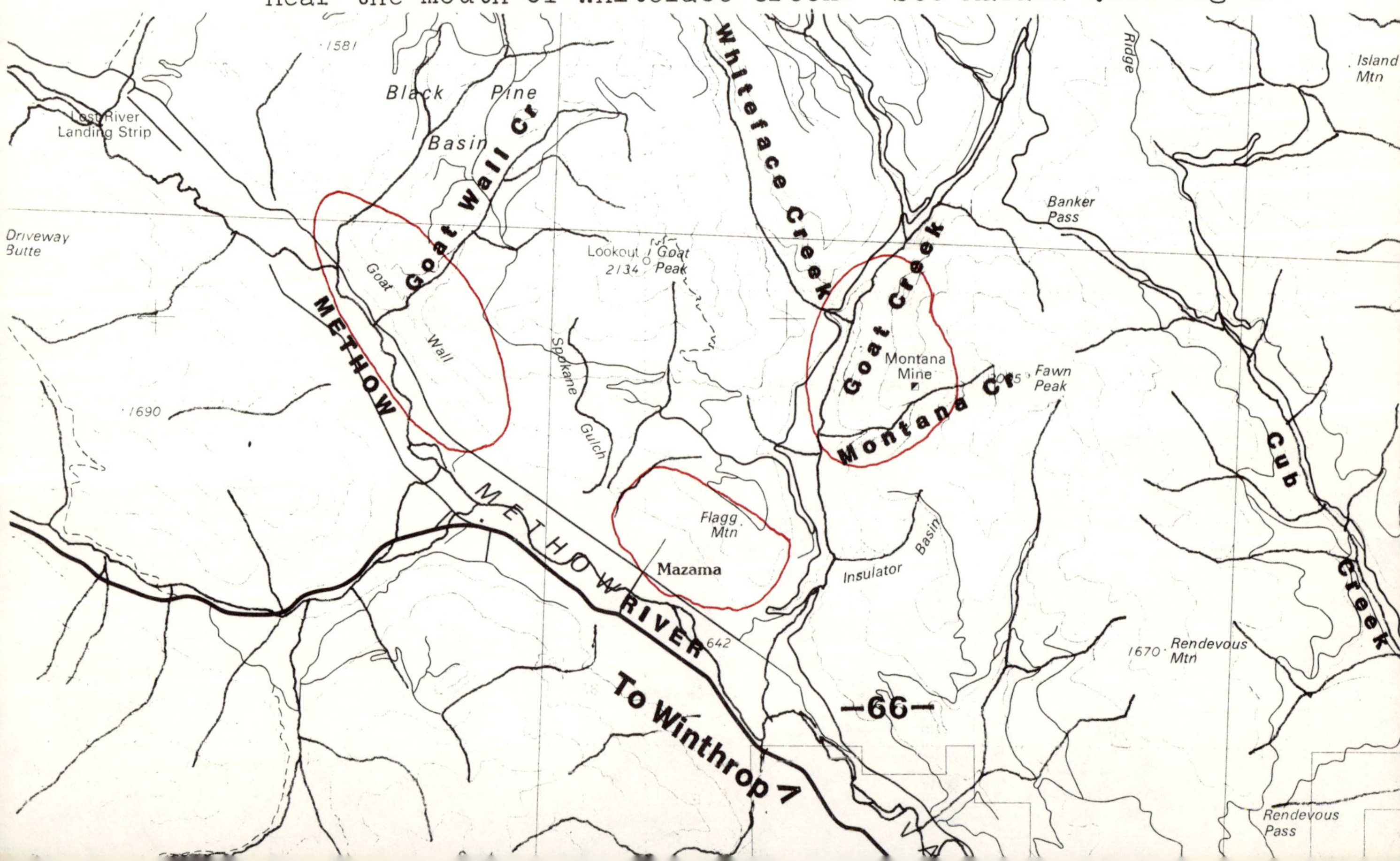

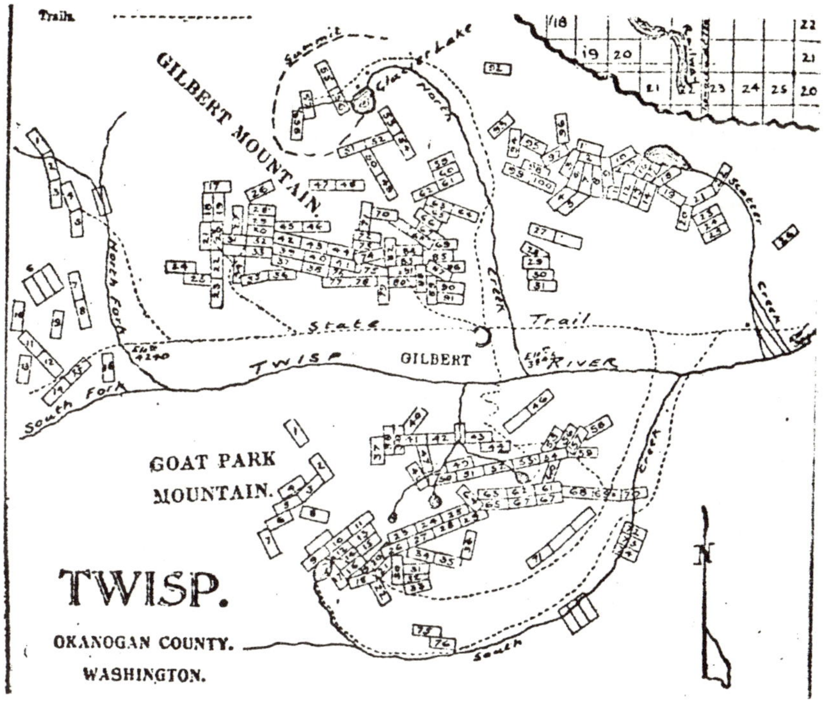

Many claims have been located in the Gilbert Mountain area near the head of the Twisp River. These mines had high gold assays in the surface ores. Exploration and development work is still being carried on. North Creek, Scatter Creek, and South Creek all have lodes in their drainage to the Twisp River. All these streams should produce some placer gold. Gilbert 7½' Quadrangle shows this area.

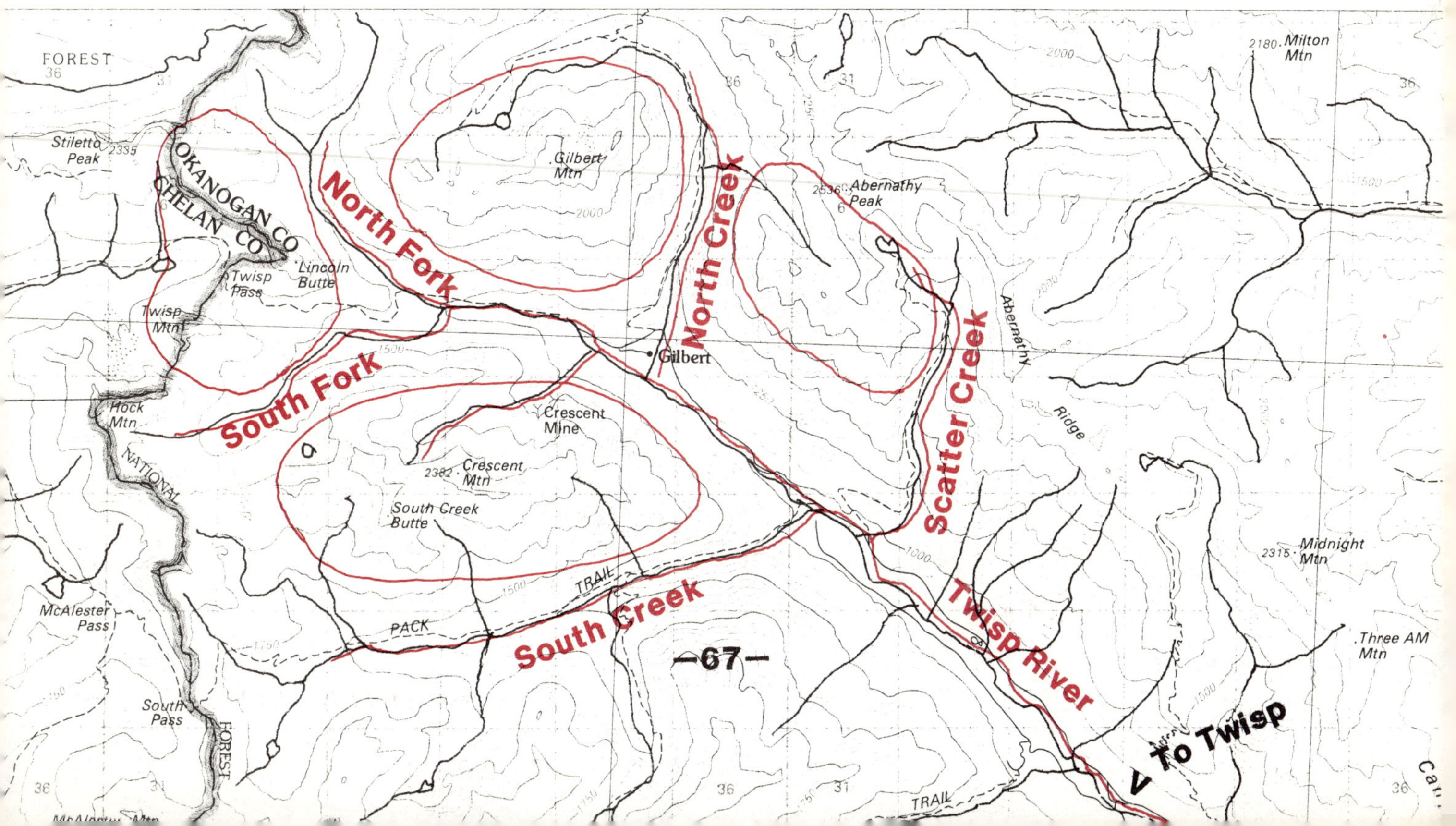

The Red Shirt Mine, east of Twisp, is still being explored. Mines in this area were discovered in 1887. These mines did not have high assay values, but have apparently been consistent enough to hold interest for almost a century.

South of Twisp, on Lookout Mountain, lode mines are also being developed at the present time. Streams flowing to the north and Smith Creek flowing south may produce some placer gold. Twisp 15' Quadrangle shows this area.

A lode mine near the head of Leecher Canyon has produced gold. Leecher Canyon is a tributary to the Methow River north of Carlton.

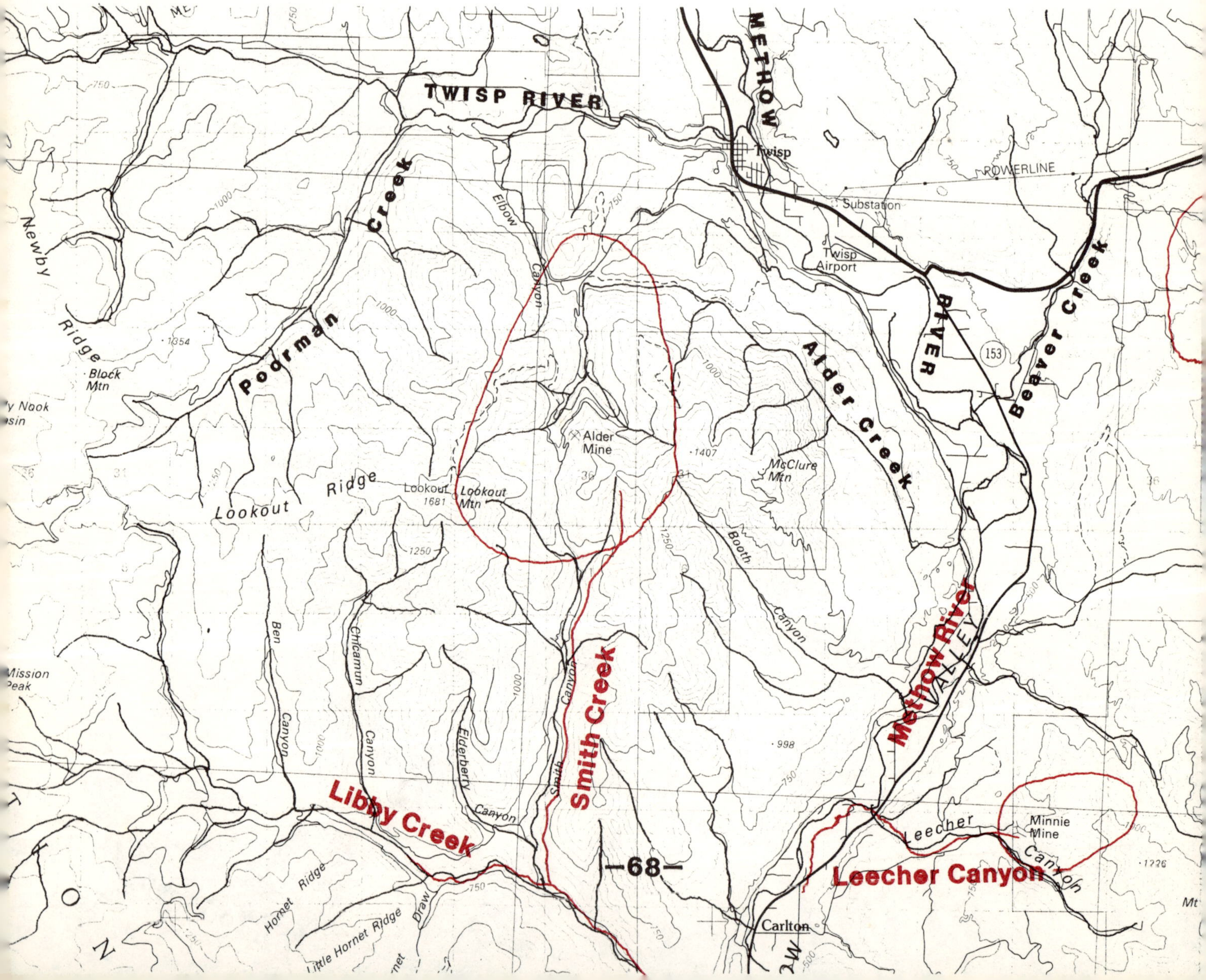

A mineralized zone near Methow begins several miles up Gold Creek and runs southeast, parallel to the Methow River and crosses the Methow River south of the town of Methow. Gold Creek, South Fork Gold Creek, McFarland Creek, Squaw Creek, and Black Canyon Creek all drain this area and all may produce placer gold. There are many lode mines in this area. Some are still active. The Methow River has had producing placers and the Columbia River below its mouth may produce placer gold. This area is shown on Brewster 15', Cooper Mountain, Methow, and Hungry Mountain 7½' Quadrangles.

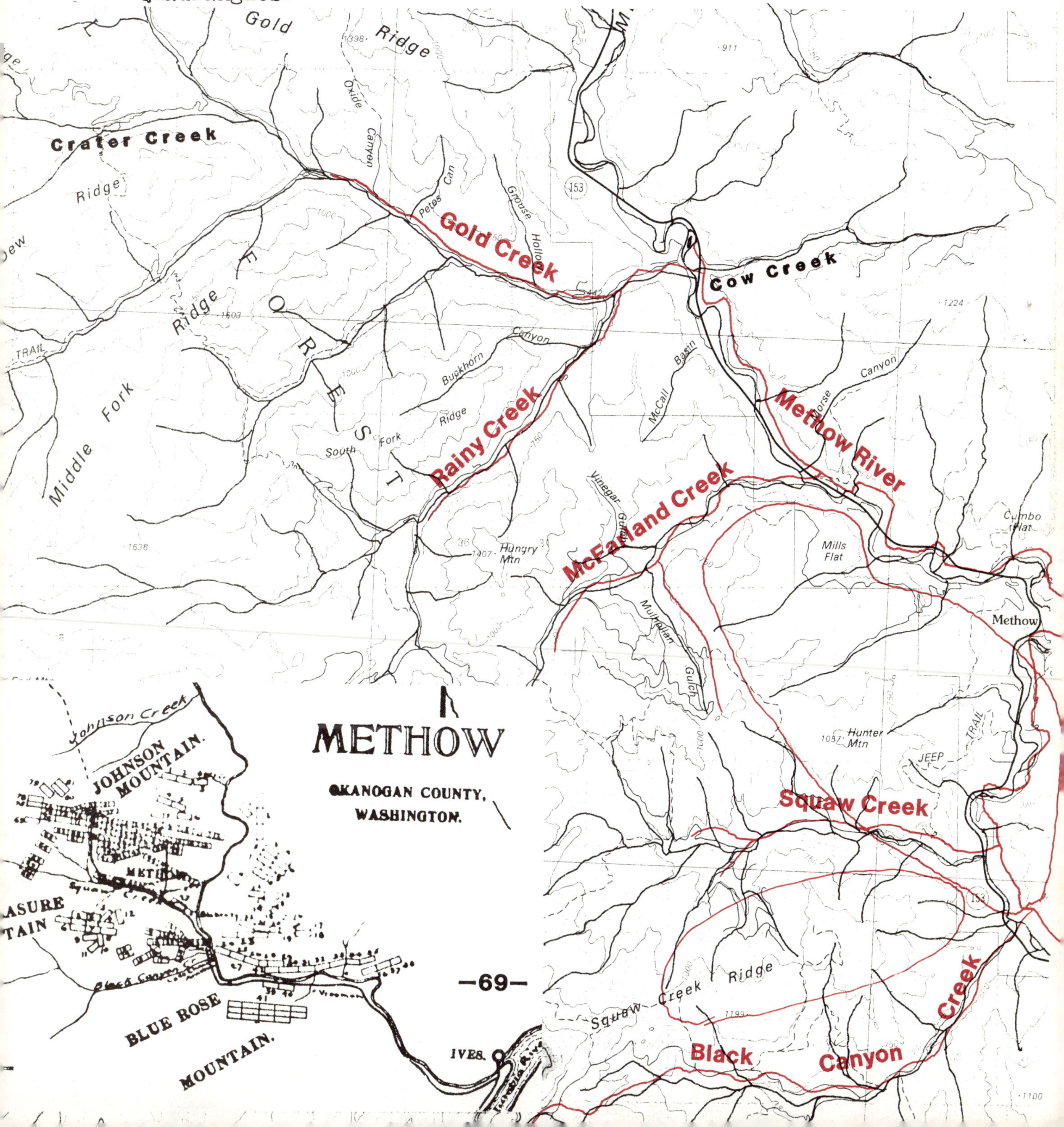

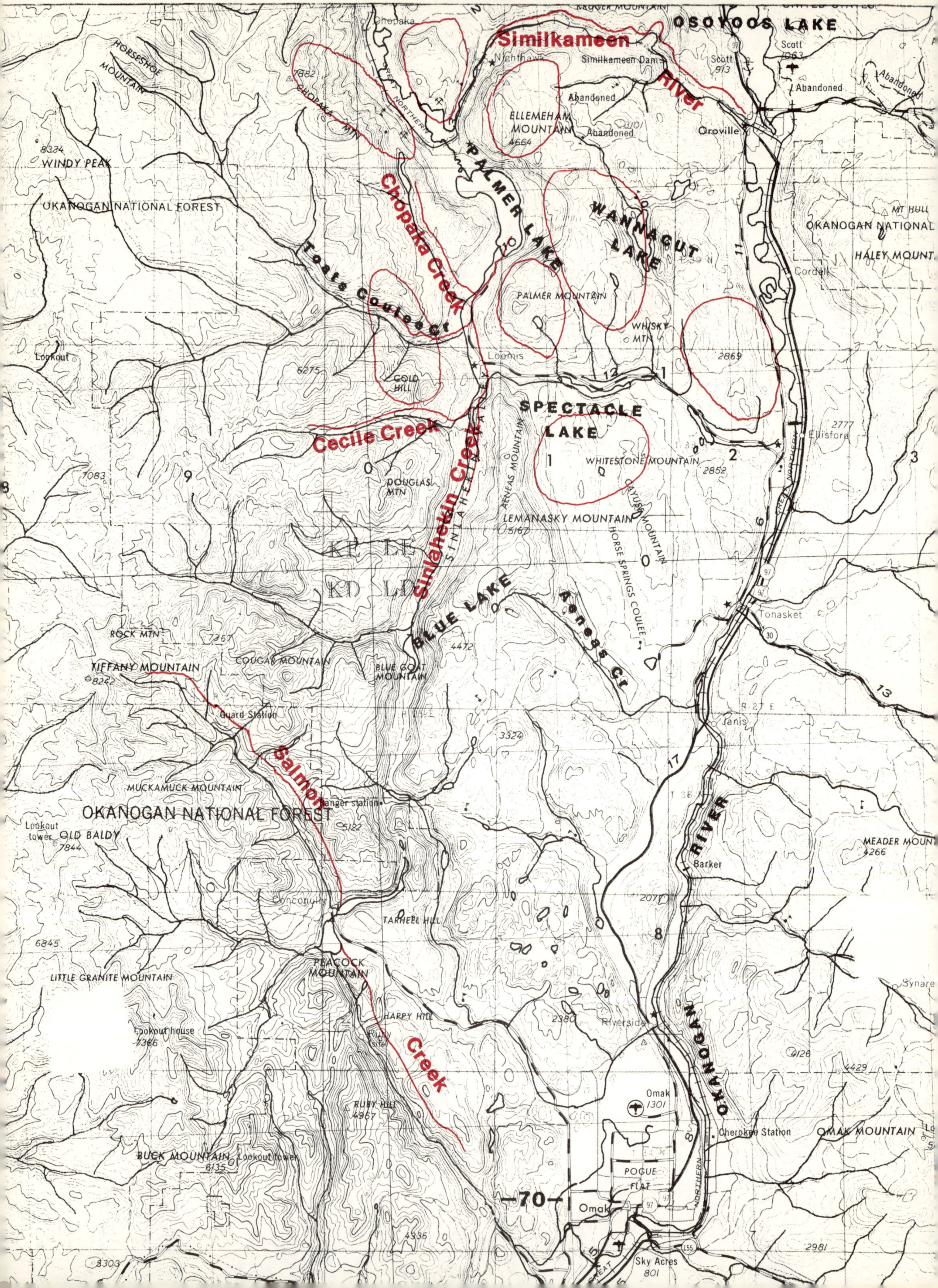

OSOYOOS LAKE
Similkameen
River
Similkameen Dam
Nighthawk
Chopaka
Scott
1063
Scott
913
Abandoned
Abandoned
Abandoned
Oroville
ELLEMEHAM
MOUNTAIN
4664
HORSESHOE
MOUNTAIN
CHOPAKA MTN
7882
8334
WINDY PEAK
OKANOGAN NATIONAL FOREST
Chopaka Creek
PALMER LAKE
WANNACUT
LAKE
MT HULL
OKANOGAN NATIONAL
HALEY MOUNT
Cordell
Toats Coulee Cr
PALMER MOUNTAIN
WHISKY
MTN
2869
Loomis
Lookout
6275
GOLD
HILL
SPECTACLE
LAKE
Cecile Creek
2777
Ellisford
WHITESTONE MOUNTAIN
2852
7083
DOUGLAS
MTN
Sinlahekin Creek
LEMANASKY MOUNTAIN
5167
HORSE SPRINGS COULEE
BLUE LAKE
Aeneas Cr
Tonasket
ROCK MTN
7257
TIFFANY MOUNTAIN
8242
COUGAR MOUNTAIN
BLUE GOAT
MOUNTAIN
4472
Guard Station
Janis
3324
Salmon
MUCKAMUCK MOUNTAIN
Ranger station
OKANOGAN NATIONAL FOREST
Lookout
tower
OLD BALDY
7844
5122
RIVER
Barker
MEADER MOUNT
4266
Conconully
TARHEEL HILL
2070
6845
PEACOCK
MOUNTAIN
LITTLE GRANITE MOUNTAIN
Synarep
HAPPY HILL
2380
Riverside
Lookout house
7366
Creek
4126
4429
OKANOGAN
Omak
1301
RUBY HILL
4967
Cherokee Station
OMAK MOUNTAIN
BUCK MOUNTAIN
6135
Lookout tower
POGUE
FLAT
70
Omak
4336
8303
Sky Acres
801
2981

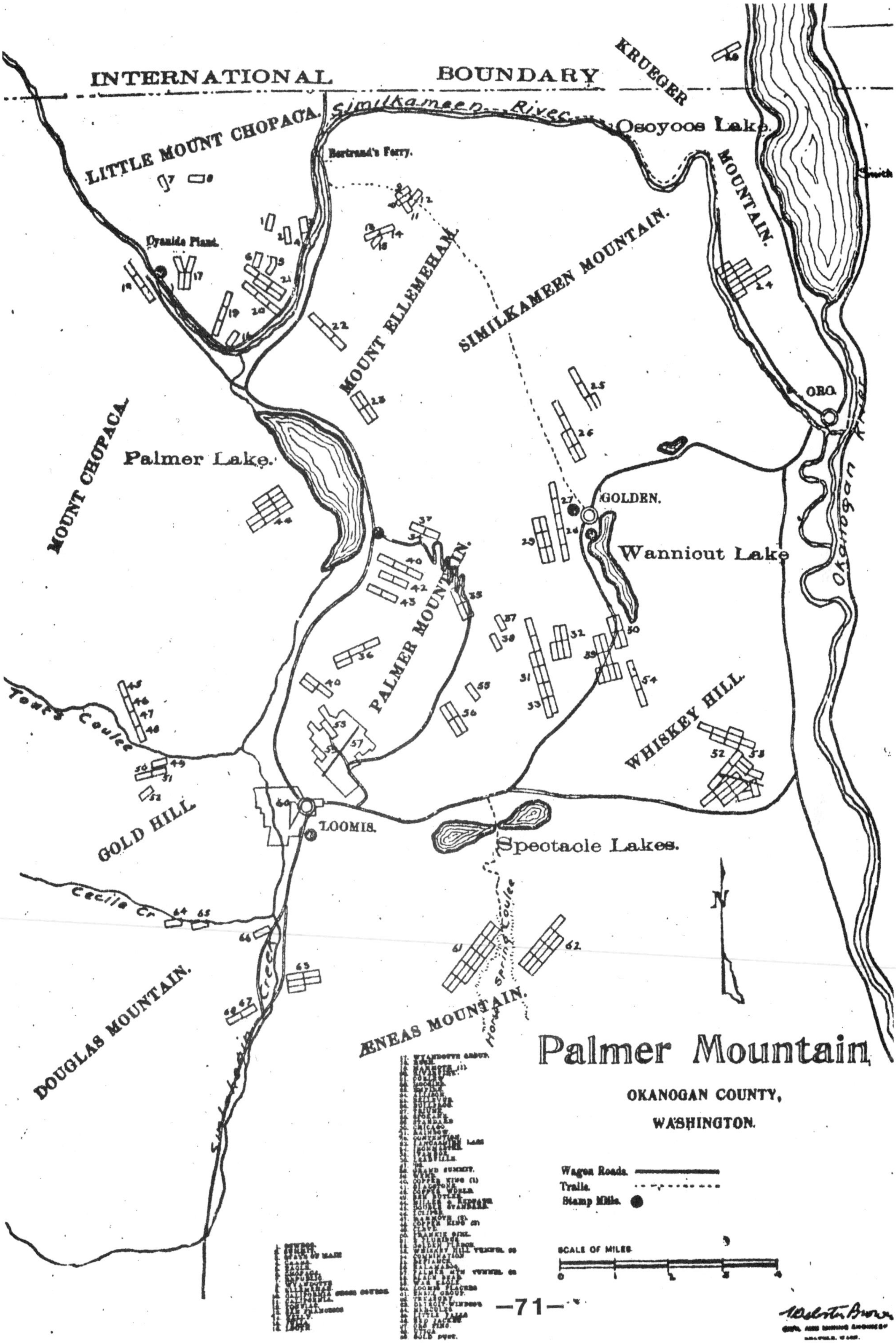
INTERNATIONAL BOUNDARY
KRUEGER MOUNTAIN
Similkameen River
Osoyoos Lake
LITTLE MOUNT CHOPACA
Bertrand's Ferry
Cyanide Plant
MOUNT ELLEMEHAM
SIMILKAMEEN MOUNTAIN
ORO
MOUNT CHOPACA
Palmer Lake
GOLDEN
Wanniout Lake
Okanogan River
PALMER MOUNTAIN
WHISKEY HILL
Tonts Coulee
GOLD HILL
LOOMIS
Spectacle Lakes
N
Cecile Cr.
Sinlahekin Creek
Spring Coulee
DOUGLAS MOUNTAIN
ÆNEAS MOUNTAIN
Palmer Mountain
OKANOGAN COUNTY,
WASHINGTON
Wagon Roads
Trails
Stamp Mills
SCALE OF MILES
0
1
2
3
4

Placer gold has been found on Salmon Creek from eight miles above Conconully to a few miles below. There are many lode mines in this area, but gold is usually a minor value. Tiffany Mountain Quadrangle shows this district of Okanogan County.

A highly mineralized area starts near Loomis and extends north to the Canadian border. This district includes the first placer gold rush and the first lode gold discovery in Washington State. The early miners found some very rich placers. Pockets of ore in lode mines had assays of unbelievalbe value. The Grand Summit Mine on Palmer Mountain assayed $39,000.00 per ton of ore. That is almost 1900 ounces per ton or $750,000.00 at today's prices.

The placer mines produced well for a time, then were abandoned for better diggings in other locations. Most streams show evidence of placer mining. The Similkameen River was the major producer, and has shown good recoveries from Nighthawk downstream to Oroville. The decomposed quartz carrying free gold through this area should replenish the placers each year on most streams. Loomis and Oroville 15' Quadrangles show this part of Okanogan County.

EASTERN WASHINGTON

Of the many mines in Eastern Washington only a few can be considered gold mines. Some gold has been found in most mines, but only as a minor value. Silver, lead copper, and zinc are the primary minerals in these mines.

An area west of Nespelem has had some production of gold. These mines have been worked for silver, with gold as a secondary value. These mines are located on Mineral Ridge and across Panama Canyon to Hobo Hill. See Alameda Flat and Nespelem Quadrangles for this area of Okanogan County.

An area fifteen miles southwest of Republic has had a small amount of gold production. This is known as the Park City District. These are lead-silver mines with a little gold. Strawberry Creek has produced placer gold. Strawberry Creek flows into Gold Creek. This area is shown on Bald Knob Quadrangle.

Twelve miles northwest of Republic, lode mines on the southwest slopes of Horseshoe Mountain have produced gold. Cougar Creek, East Fork Cougar Creek, and Toroda Creek have produced placer gold. Aeneas and Bodie Mountain Quadrangles show this area.

Quartz and calcite veins near Bodie have produced gold from several thousand feet of tunnels. Bodie Creek and Toroda Creek may show some placer gold. Bodie Mountain Quadrangle shows this area.

The Myers Creek District covers a large area of both lode and placer mines. Lodes have been located on Bimetalic Mountain at the head of Myers Creek and on Strawberry Mountain at the head of North Fork Myers Creek. Several mines have been worked in the vicinity of Chesaw.

Mary Ann Creek and North Fork Mary Ann Creek have had good placer gold production. A clay seam four feet above bedrock carried high values a mile south of Chesaw. Deadman Creek has placers for several miles of its length. Old channels produced well on high bars two hundred and fifty feet up the mountainside. Mt. Bonaparte Quadrangle shows this area of Okanogan County.

Chesaw
Mary Ann Creek
Myers Creek
Toroda Creek
Kettle River
Curlew
Goosmus Creek
Little Goosmus Cr
Curlew Creek
CURLEW LAKE
Republic
Aeneas Cr
WEST FORK
SANPOIL RIVER
Strawberry Creek
O'Brien Cr
OKANOGAN NATIONAL FOREST
COLVILLE NATIONAL FOREST
OKANOGAN COUNTY
FERRY COUNTY
BUCKHORN MOUNTAIN
HEHE MOUNTAIN
KNOB HILL
BIMETALIC MOUNTAIN
Havillah
STRAWBERRY MOUNTAIN
SPUR MOUNTAIN
MT BONAPARTE
FOURTH OF JULY RIDGE
Wauconda
LITTLE BONAPARTE MOUNTAIN
CAYUSE MOUNTAIN
MT ANNIE
FIR MOUNTAIN
BODIE MOUNTAIN
HARDSCRABBLE MOUNTAIN
HORSESHOE MOUNTAIN
TORODA MOUNTAIN
IRENE MOUNTAIN
BAMBER MOUNTAIN
STORM KING MOUNTAIN
BALD PEAK
KLONDIKE MOUNTAIN
Tom Thumb Mine
Mountain Lion Mine
Knob Hill Mine
Flag Hill Mine
GIBRALTAR MOUNTAIN
COPPER MOUNTAIN
COOKE MOUNTAIN
GRANITE MOUNTAIN
MT ELIZABETH
FRANSON PEAK
LITTLE VULCAN MOUNTAIN
VULCAN MOUNTAIN
SNEED MTN
TUNK MOUNTAIN
AENEAS MOUNTAIN
BAILEY MTN
DUGOUT MOUNTAIN
STRAWBERRY MOUNTAIN
LITTLE MOSES MOUNTAIN
BARRET BUTTE
BALD KNOB
CAMEL BACK
QUARTZ MTN
EAGLE ROCK
BROWN MOUNTAIN
IRON MOUNTAIN
COUGAR MOUNTAIN
BEAR MOUNTAIN
THIRTEENMILE MOUNTAIN
SEVENTEENMILE MOUNTAIN

In the Curlew area of Ferry County a number of mines have produced copper and gold ore. Early miners located many mines in this district, some with very good assay values. Placers have been worked in numerous locations along the Kettle River. Big Goosmus Creek showed good placer values and has had recent production. The many big boulders make mining difficult in this narrow creek bottom. See Curlew Quadrangle.

The Republic Mining District, made famous by the fabulous Knob Hill Mine, has been a consistent producer of gold since the early 1900's. Knob Hill was one of the leading producers of gold in the United States for several years. Most of the mines in this area are concentrated in an area six miles long and four miles wide. See Republic and Aeneas Quadrangles.

Some streams draining Gold Hill between Kettle Falls and Marcus may produce some placer gold. Kettle Falls and Marcus 7½' Quadrangles show this area.

The First Thought Mine near Orient has been a major producer of lode gold. This mine produced for almost forty years on ore averaging one half ounce per ton, with high values in silver and copper. Oreint 7½' Quadrangle shows this area.

Adjacent to the First Thought Mine claims, an area known in the early days as the Pierre Lake District covers a large area between the Kettle River and the Columbia River. Many mines had high gold values in the surface outcrops, but became more valuable for other metals as the mines were worked to greater depth.

The Kettle River has been placered for much of its length through this area. Toulou Creek above and below Pierre Lake has produced placer gold. Many of the streams flowing into the Columbia River have produced placer gold and some of these should replenish each year. Laurier, Churchill Mountain, Belshazzar Mountain, Northport, Boundary, China Bend, and Bossburg 7½' Quadrangles cover this area.

From the vicinity of Kettle Falls, a rich mineral belt extends north to the Canadian border. Many of these mines carry gold as a secondary value. Almost every bar on the Columbia River north of here has had production of placer gold. These bars are all flooded under many feet of water at this time.

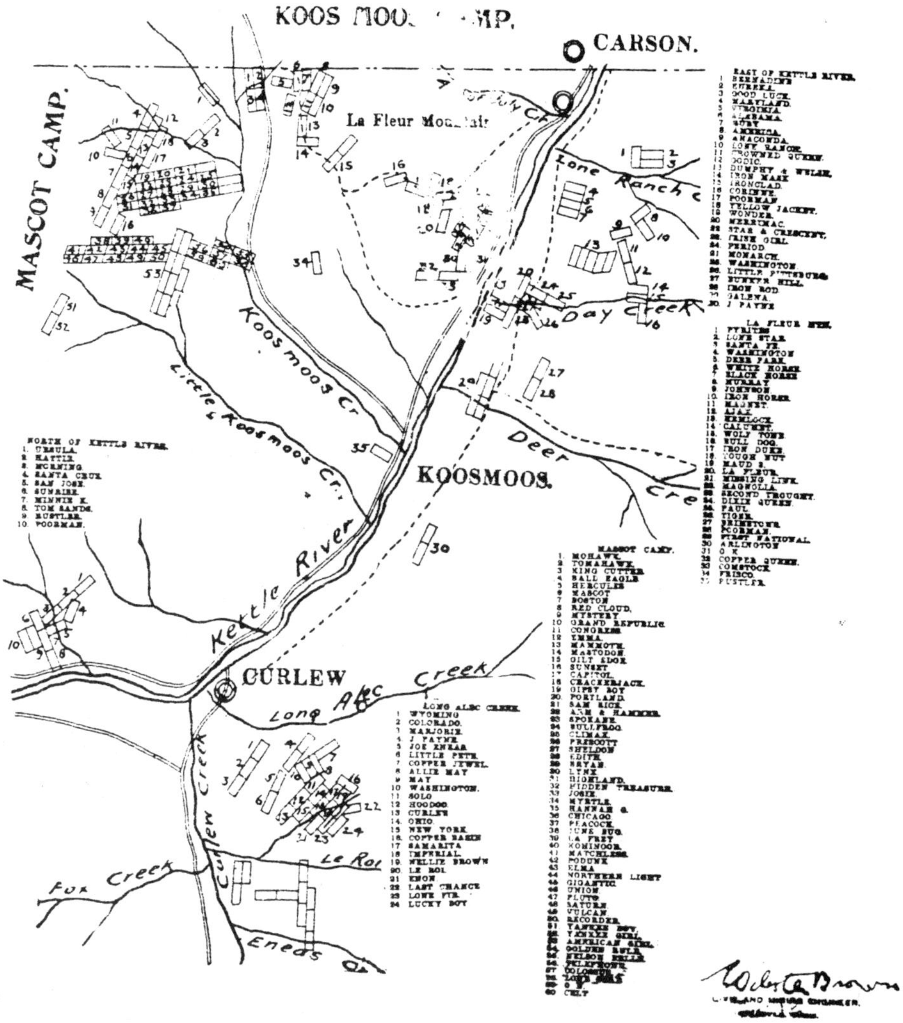
CARSON.
MASCOT CAMP.
La Fleur Mountain
Lone Ranch C
Day Creek
Koosmoos Cr
Little Koosmoos Cr.
Deer Cre
KOOSMOOS.
Kettle River
CURLEW
Long Alec Creek
Curlew Creek
Fox Creek
Le Roi
Eneas Cr.
NORTH OF KETTLE RIVER.
MASCOT CAMP.
LONG ALEC CREEK
LA FLEUR MTN.
EAST OF KETTLE RIVER.

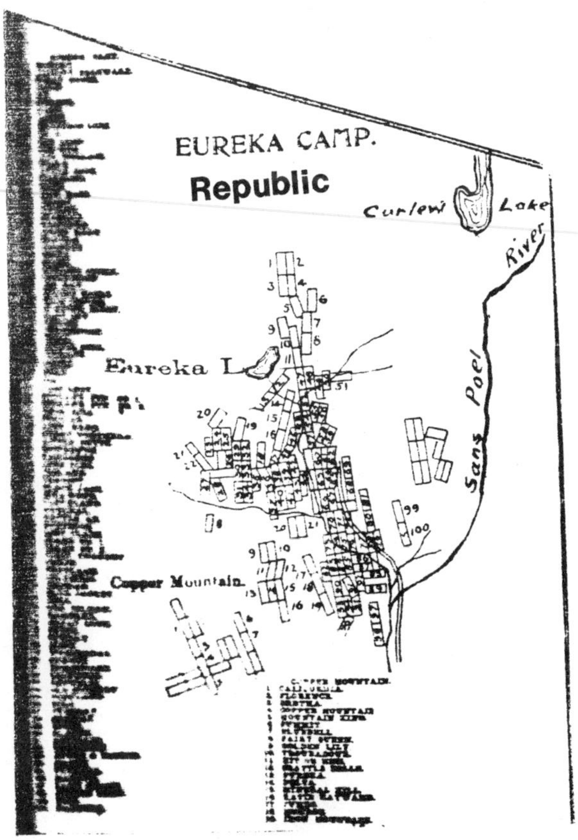
EUREKA CAMP.
Republic
Curlew Lake
River
Eureka L
Sans Poel
Copper Mountain
COPPER MOUNTAIN.

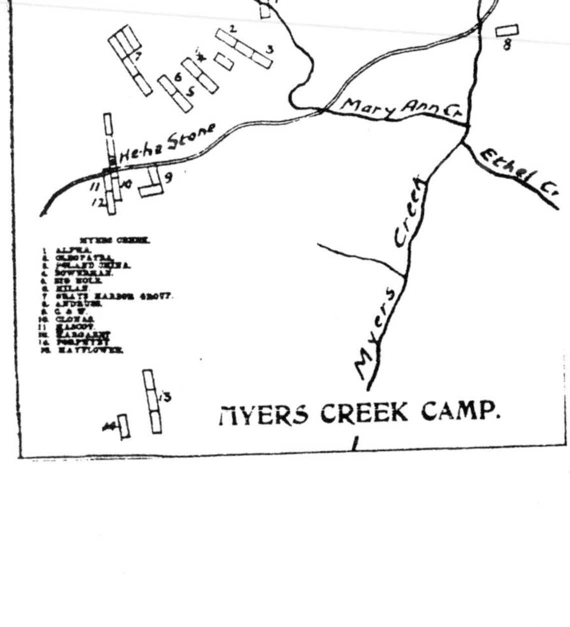
Mary Ann Cr
He-ha Stone
Ethel Cr
Myers Creek
MYERS CREEK CAMP.

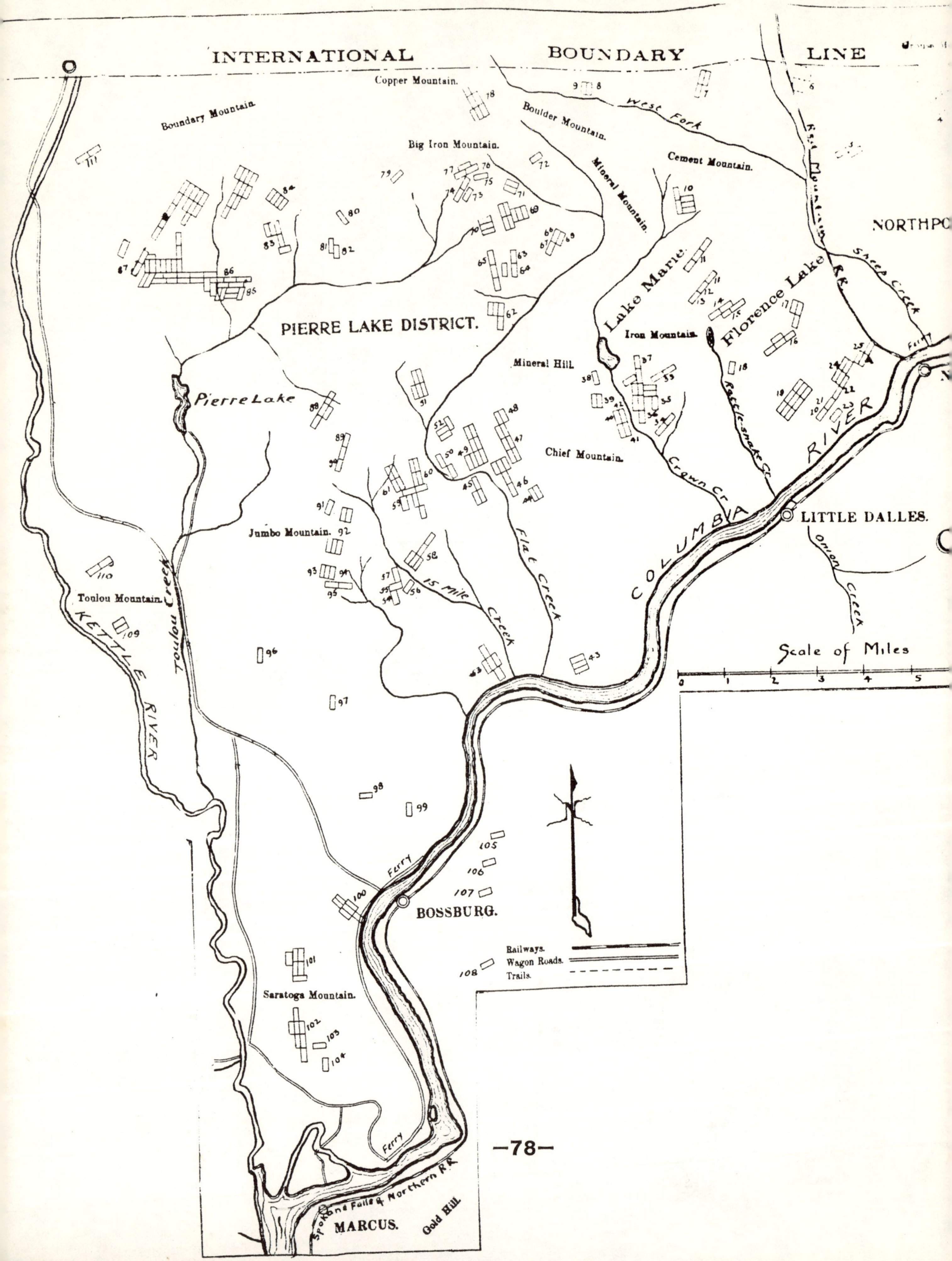
INTERNATIONAL BOUNDARY LINE
Copper Mountain.
Boundary Mountain
Big Iron Mountain.
Boulder Mountain.
West Fork
Mineral Mountain
Cement Mountain.
NORTHPO
PIERRE LAKE DISTRICT.
Lake Marie
Iron Mountain
Florence Lake
Sheep Creek
Mineral Hill.
Pierre Lake
Chief Mountain.
Crown Cr
Rattle-snake Cr.
Columbia River
LITTLE DALLES.
Onion Creek
Jumbo Mountain.
15 Mile Creek
Flat Creek
Toulou Mountain.
Toulou Creek
Kettle River
Scale of Miles
0 1 2 3 4 5
Ferry
BOSSBURG.
Railways.
Wagon Roads.
Trails.
Saratoga Mountain.
Spokane Falls & Northern R.R.
MARCUS.
Gold Hill

Near Inchelium several lode mines have produced gold along with other minerals. In most mines silver and lead have the major values with gold as a by-product. Placers along the Columbia River were good producers, but are now flooded by the waters impounded by Grand Coulee Dam. A good placer was worked near the mouth of Stranger Creek. Other streams in the area may produce a few colors. See Inchelium Quadrangle.

Copper mines near Keller have produced a very limited amount of gold. See Keller Quadrangle.

Several mines in the southern half of Stevens County have produced minor amounts of gold in the lead, silver, and copper deposits. The United Copper Mine, near Chewelah, produced 1,300 ounces of gold along with 9,714,504 pounds of copper over a fifty year period of time. Chewelah Quadrangle shows this area.

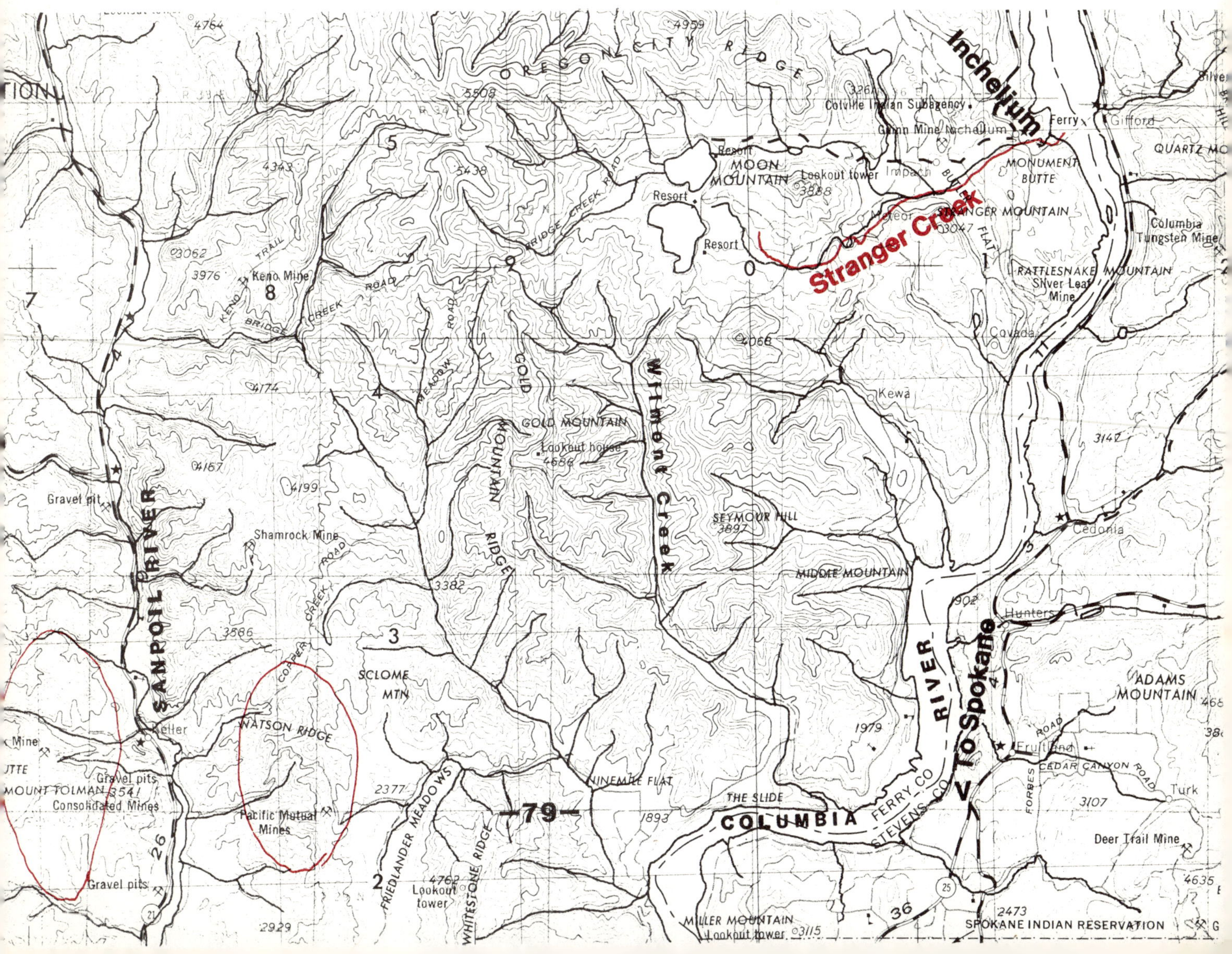

Lead mines near Metaline Falls have produced a minor amount of gold. Lode gold has been produced near Gypsy Meadows, near the head of Sullivan Creek. Some placer gold has been recovered from Sullivan Creek.

Placer gold has been found on the Pend Oreille River below Zee Canyon and downstream for several miles. Boundary Dam, Metaline Falls, and Pass Creek Quadrangles show this area of Pend Oreille County.

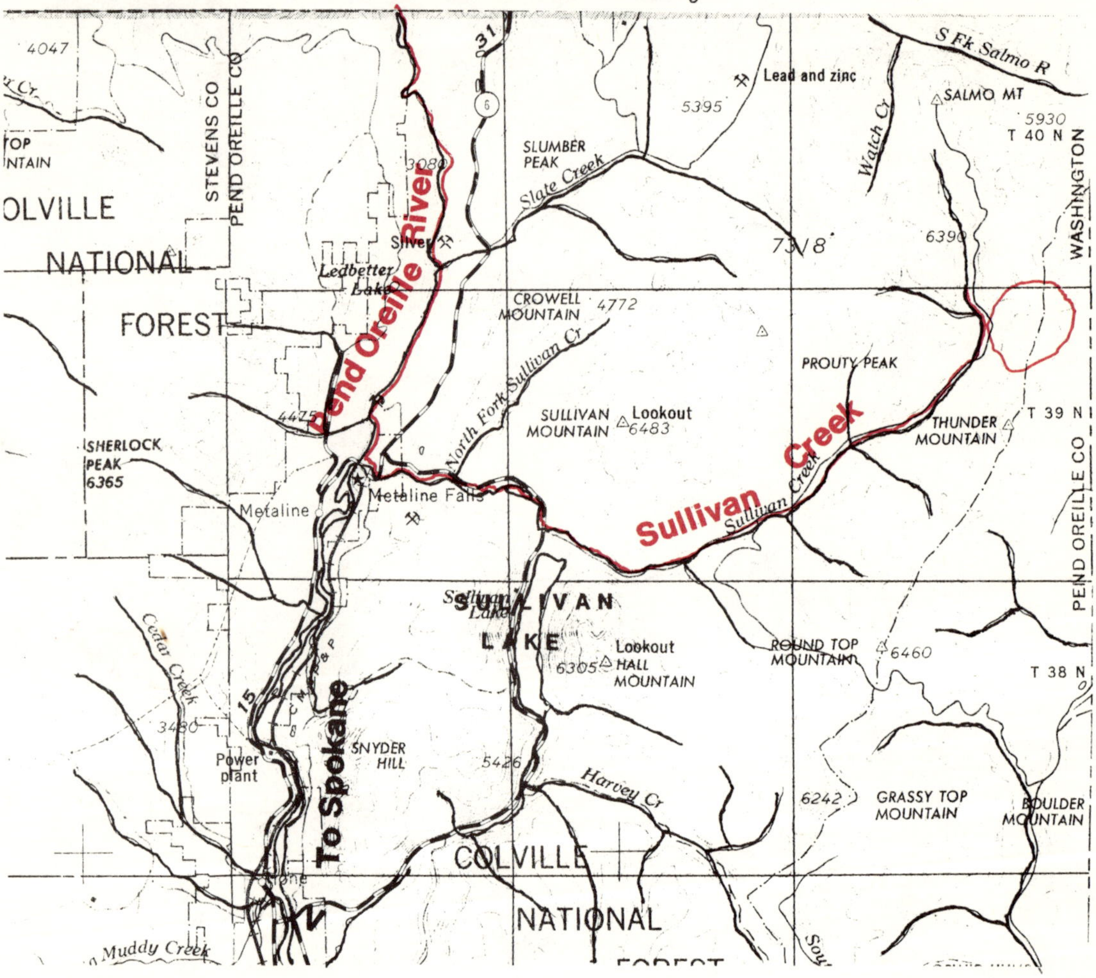

The southeast corner of Washington State has had very limited production of gold. The Palouse River between Palouse and Colfax may produce a few colors of very fine flour gold. The Snake River has had a few placer operations on sand bars, but these are now flooded. A better bet might be a search for the Lost Shovel Creek Mine somewhere in Asotin County.

RECOMMENDED READING

FOR OVER TEN YEARS, Ram books have helped hobby and professional metal detector users LEARN HOW TO BE SUCCESSFUL treasure hunters and electronic prospectors. If you are new to metal detector use and want the most from your instrument, you should study their excellent, easy-to-understand how-to books written by experts in the field.

* ELECTRONIC PROSPECTING. Garrett, Lagal, Grant. 96 pp. $4.95. Learn to find gold/silver veins/pockets/nuggets using easy electronic metal detector methods.

* GOLD PANNING IS EASY. Lagal. 112 pp. $5.95. This excellent field guide tells you how to find and pan gold easily and quickly.

VLF-TR METAL DETECTOR HANDBOOK. Lagal, Garrett. 200 pp. $7.95. The operations/technical manual. VLF/TR detectors and use explained and compared with other types.

DETECTOR OWNER'S FIELD MANUAL. Lagal. 236 pp. $8.95. Complete field guide. Explains capabilities/how-to-use procedures for all types of metal detectors.

PROFESSIONAL TREASURE HUNTER. Mroczkowski. 154 pp. $6.95. Discover how to succeed with professional methods, persistence, and hard work.

ROBERT MARX: QUEST FOR TREASURE. Marx. 286. pp. $9.95. The exciting true account of the discovery and salvage of a Spanish treasure galleon lost sunk in 1656.

SUCCESSFUL COIN HUNTING. Garrett. 248 pp. $7.95. The classic. authoritative guide to finding valuable coins with all types of metal detectors.

TREASURE HUNTER'S MANUAL #6. von Mueller. 318 pp. $9.95. The standard to guide inexperienced beginners through the mysteries of full time treasure hunting.

TREASURE HUNTER'S MANUAL #7. von Mueller. 334 pp. $9.95. Another classic.... THE book on professional methods of research, recovery, and disposition of treasures found.

TREASURE HUNTING PAYS OFF! Garrett. 92 pp. $3.95. An excellent introduction to all facets of treasure hunting.

These books may be purchased from dealers listed on the following pages or ordered from Ram Books, P.O. Drawer 38649, Dallas, TX 75238.

MINING AND PROSPECTING SUPPLIES AND EQUIPMENT

International Prospectors Supply
Division of Pearl Electronics
312 Dexter North, Seattle, WA 98109
Telephone 206-622-6200

Gold Pans-Dredges-Sluices-Metal Detectors-Specialized Equipment
Books-Maps Large stock and Rentals

R&G Enterprises
16111 OK Road
Snohomish, WA 98290
Telephone 206-334-6692

Gold Pans-Dredges-Sluices-Metal Detectors-Books-Maps

Long's Prospecting & Treasure Hunting Supplies
2416½ East Main
Lewiston, Idaho 83501
Telephone 208-743-1768

Gold Pans-Dredges-Sluices-Metal Detectors-Specialized Equipment
Books-Maps

Gold Rush Mining Supplies
4242 Van Giesen
West Richland, WA 99352
Telephone 509-967-3307

Gold Pans-Dredges-Sluices-Metal Detectors-Specialized Equipment
Books-Maps-Topographic Maps

D&K Detector Sales Inc.
13809 SE Division
Portland, OR 97236
Telephone 503-761-1521

Gold Pans-Dredges-Sluices-Metal Detectors-Specialized Equipment
Books-Maps-Topographic Maps

Northwest Territorial Mining & Mfg. Co.
28207 85th Ave. South
Kent, WA 98031
Telephone 206-859-1446

Manufactures and Distributes gold recovery systems.
Square Gold Pans-Classifiers-Dredges-Sluices-Books

Renton Coin and Detectors
225 Wells Ave. South
Renton, WA 98055
Telephone 206-226-3890

Gold Pans-Sluices-Metal Detectors-Books-Maps

MINING AND PROSPECTING SUPPLIES AND EQUIPMENT

Miners Incorporated
P.O. Box 11301
Riggins, Idaho 83549
Telephone 208-628-3247

Gold Pans-Dredges-Sluices-Metal Detectors-Specialized Equipment
Books- Maps
Miners offer a large inventory of mining specialties and books. Many are not available from any other source. Free Catalog.

Western Sports
S. 3725 Stone
Spokane, WA 99203
Telephone 509-448-8686

Manufactures Dredges and Sluices

Bowens Hideout
S. 1823 Mt. Vernon
Spokane, WA 99203
Telephone 509-534-4004

Gold Pans-Dredges-Sluices-Metal Detectors-Specialized Equipment
Books-Maps Catalog($2.00 Shipping Charge)

Doug's Mining Supplies
14320 NE 20th Street
Bellevue, WA 98007
Telephone 206-746-0227

Gold Pans-Dredges-Sluices-Metal Detectors-Specialized Equipment
Books-Maps

The Shorey Book Store
110 Union Street
Seattle, WA 98101
Telephone 206-624-0221

Books and Maps. Many rare books and reprints on mining and prospecting. Free catalog "Shorey Publications".

Metsker Maps of Seattle
702 First Avenue
Seattle, WA 98104
Telephone 206-623-8747

Topographic Maps-Property Ownership Maps-State & County Maps

Wenatchee Prospecting Supplies
26 N Wenatchee Ave.
Wenatchee, WA 98801
Telephone 509-662-3780

Gold Pans-Dredges-Sluices-Metal Detectors-Specialized Equipment
Books-Maps

NUGGET NOTES

Date	Page	Remarks

NUGGET NOTES

Date	Page	Remarks

NUGGET NOTES

Date	Page	Remarks